Praise for Debbie Miller's *Midnight Wilderness*:

"Miller, a freelance writer and photographer, presents a timely plea for the preservation of the largest wilderness area in the United States, now threatened by oil and gas development. . . . She describes vividly the wonders of this magnificent 19-million-acre preserve in Alaska's northeastern corner."

—*Publishers Weekly*

"Miller's prose is lively and evocative, her perspective is insightful, her relationship to this arctic wildland is intense and inspiring."

—Richard Nelson, author of *The Island Within* and the Alaska State Writer

"Miller transmits Alaska's magnificence to the reader through vivid descriptions of immense vistas and fragile wildlife."

—*Booklist*

"Miller's trips to the refuge are inspiring to those who dream of being truly alone and self-sufficient in the wild. . . ."

—*Tampa Tribune Times*

"Debbie Miller's book, *Midnight Wilderness*, describes this large, isolated wilderness with lyric precision. It is an intimate, knowledgeable work. . . . Former president Jimmy Carter read the book and was inspired to meet the author, visiting their camp during the Millers' most recent trip to the refuge."

—*San Francisco Examiner*

"There can be few people who know this troubled and lovely land as well as Debbie Miller, and it is that gift of knowledge she brings with a seeing eye and feeling heart."

—*Wilderness Magazine*

"*Midnight Wilderness* is an unparalleled look at an area few Alaskans, and still fewer visitors, will ever see. The glimpse it provides makes us long to see the refuge for ourselves, and to hope it doesn't change before we get the chance."

—*Homer News*

"On foot and in a kayak, Miller explored more than 1,000 miles of this section of northeastern Alaska that's being threatened by oil exploration. Her book offers glimpses in the area's wilderness, wildlife, and recreational values that must be preserved for future generations."

—*Backpacker* magazine

"Debbie Miller's account of her 1,000-mile exploration of Alaska's Arctic National Wildlife Refuge is sure to grab the adventurer in all of us."

—*San Diego Tribune*

"There is some first-rate personal narrative throughout *Midnight Wilderness*. . . . Miller brings the country to life—people and place, as well as politics and principles."

—*Bloomsbury Review*

"Miller has visited this wildlife sanctuary over a period of thirteen years. A naturalist, she witnesses wilderness firsthand. She takes time to appreciate what is there."

—*The Anchorage Times*

"After reading *Midnight Wilderness*, one is not only entertained but also extremely well-informed about this ecosystem on which millions of lives depend, including thousands of human lives. Her work is reminiscent of the nature writing of U.S. Supreme Court Justice William O. Douglas in noting all the interesting floral and wildlife events along a day's march. . . . *Midnight Wilderness* encourages reflection on the meaning of Earth itself, and in a lovely preface, Margaret Murie joins Miller in calling for the courage to protect the region, not only for future humanity but also 'for the sake of the land itself . . . empty of technology and full of life.'"

—John M. Kauffmann, in *Alaska's Brooks Range*

"*Midnight Wilderness* is an eloquent argument for preservation; the author's love for this austere but beautiful realm is apparent on every page."

—John A. Murray, in *Western American Literature*

"Miller did not write these pages under the gun of current events, but rather under the charm of days spent wandering under a heavy pack across the Coastal Plain and through the Brooks Range. . . . She writes: 'This is truly the most extraordinary of ancient birthplaces. . . . There is something inherently special about places where humans and all members of the animal kingdom bear their young. Such places are looked to with reverence, renewal of spirit, and celebration of life . . . .' There is much to savor in *Midnight Wilderness*."

—*Fairbanks Daily News-Miner*

"*Midnight Wilderness* is factually and beautifully written. Readers share the grandeur of Alaska's Arctic National Wildlife Refuge. . . ."

—*Bridgeport Post* (Connecticut)

Midnight Wilderness

The moon rises over Arctic Village.

Midnight Wilderness

Journeys in Alaska's Arctic National Wildlife Refuge

Debbie S. Miller

Foreword by Margaret E. Murie

First edition, 1990. Second edition, 2000. Third edition, 2011.

Manufactured in the United States of America

Cover and Book Design: Jane Jeszeck
Layout: Emily Brooks Ford
Cartographer: Victoria Hand

All photographs by Debbie S. Miller unless otherwise noted

Cover photograph © Michio Hoshino/Minden Pictures/National Geographic Stock

Library of Congress Cataloging-in-Publication Data

Miller, Debbie S.
Midnight wilderness : journeys in Alaska's Arctic National Wildlife Refuge / Debbie S. Miller ; foreword by Margaret E. Murie.
p. cm.
Originally published: San Francisco : Sierra Club Books, c1990.
Includes bibliographical references and index.
ISBN 978-1-59485-633-4
1. Natural history—Alaska—Arctic National Wildlife Refuge.
2. Arctic National Wildlife Refuge (Alaska)—Description and travel.
3. Miller, Debbie S.—Travel—Alaska—Arctic National Wildlife Refuge.
4. Wilderness areas—Alaska. I. Title.
QH105.A4M53 2011
508.798—dc22
2011008581

A portion of the royalties from sales of *Midnight Wilderness* will be donated to the Alaska Wilderness League, an organization that works to protect and preserve Alaska's wild lands, including the Arctic National Wildlife Refuge.

ISBN (paperback): 978-1-59485-633-4
ISBN (ebook): 978-1-59485-634-1

For those who believed in the dream of
an Arctic Refuge and fought for it
and for Robin and Casey and future
generations of wilderness seekers

Contents

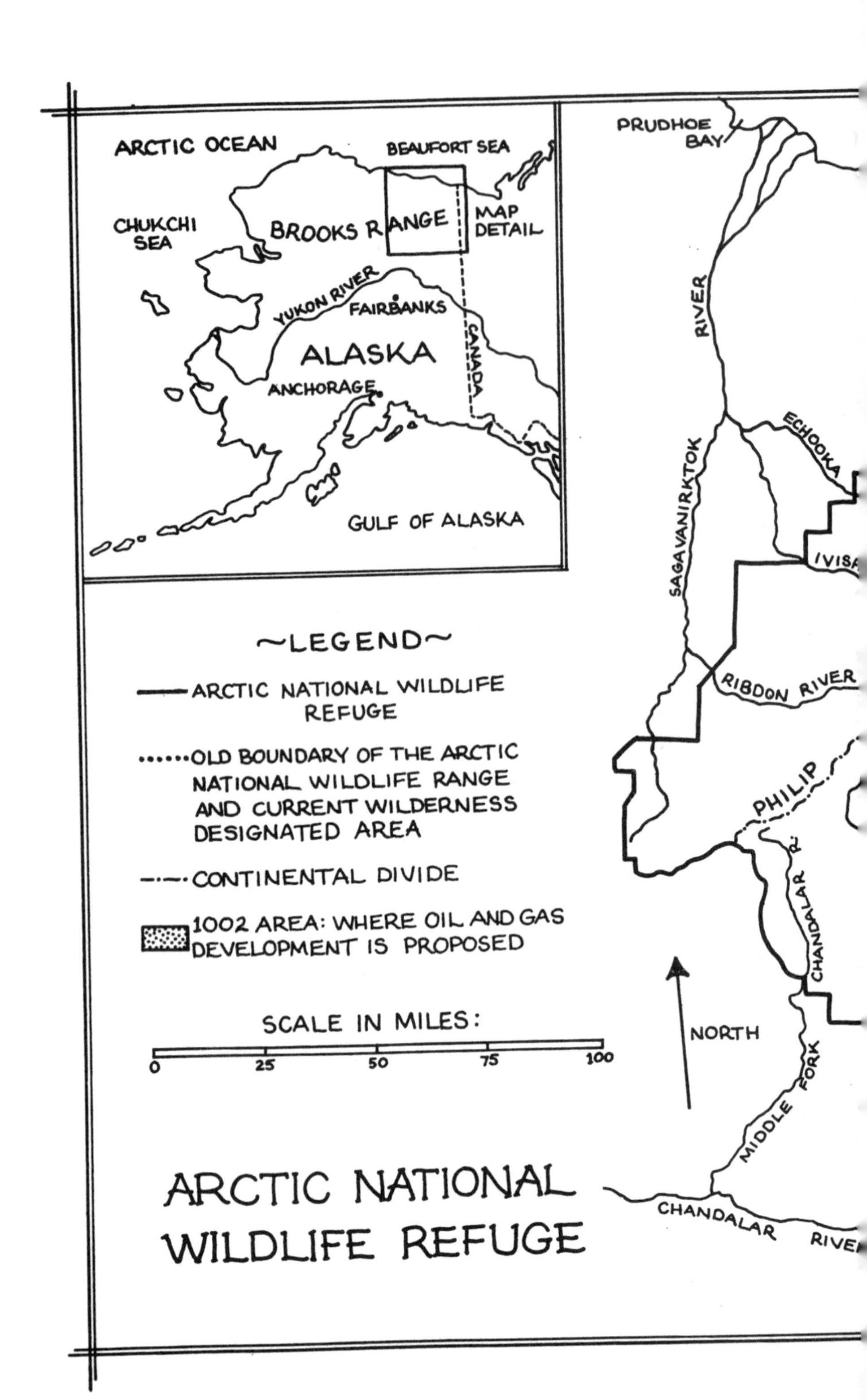
ARCTIC OCEAN
BEAUFORT SEA
CHUKCHI SEA
BROOKS RANGE
MAP DETAIL
YUKON RIVER
FAIRBANKS
ALASKA
CANADA
ANCHORAGE
GULF OF ALASKA
~LEGEND~
ARCTIC NATIONAL WILDLIFE REFUGE
OLD BOUNDARY OF THE ARCTIC NATIONAL WILDLIFE RANGE AND CURRENT WILDERNESS DESIGNATED AREA
CONTINENTAL DIVIDE
1002 AREA: WHERE OIL AND GAS DEVELOPMENT IS PROPOSED
SCALE IN MILES:
0
25
50
75
100
ARCTIC NATIONAL WILDLIFE REFUGE
PRUDHOE BAY
RIVER
SAGAVANIRKTOK
ECHOOKA
RIBDON RIVER
PHILIP
CHANDALAR R.
NORTH
MIDDLE FORK
CHANDALAR

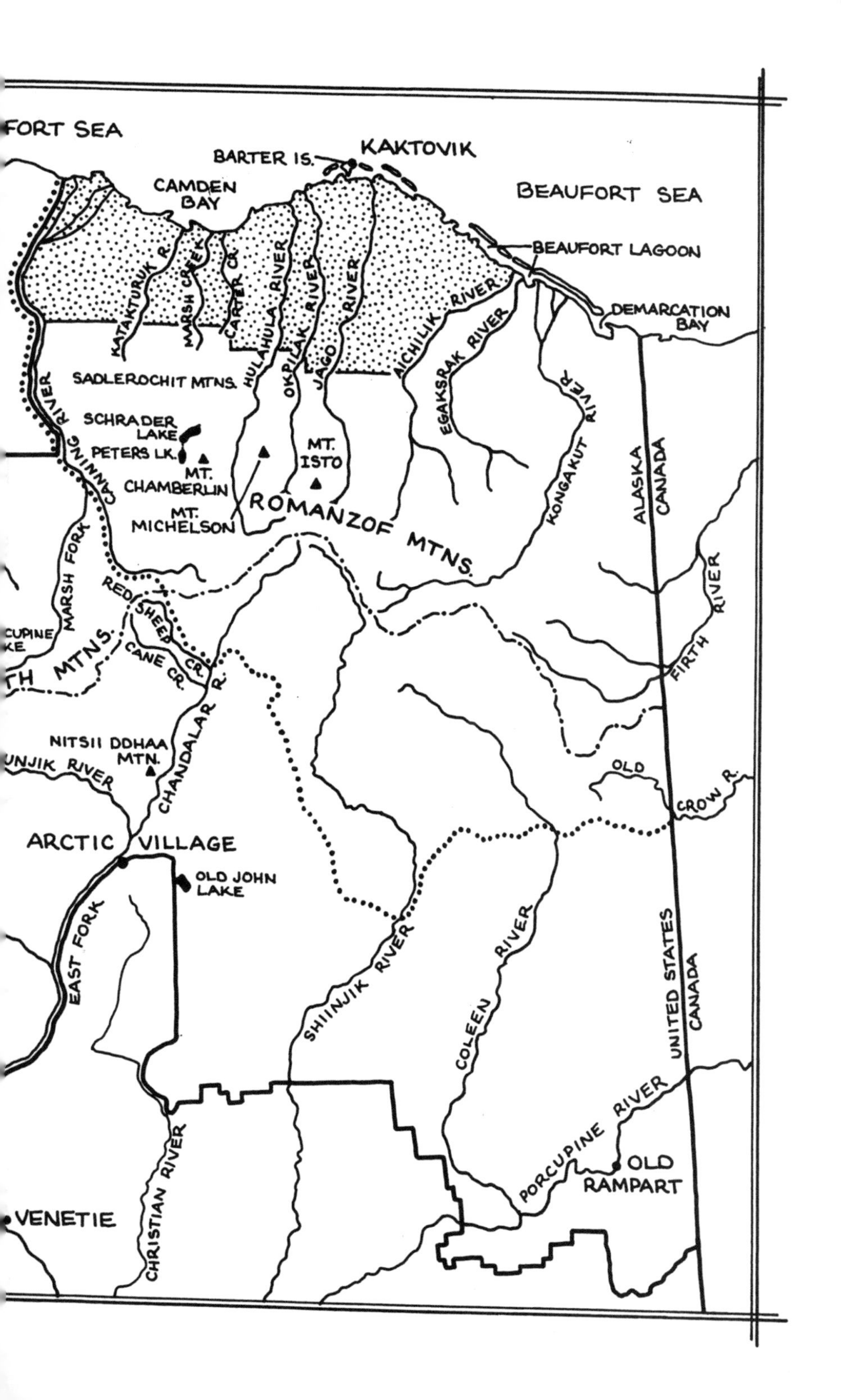

FORT SEA
BARTER IS.
KAKTOVIK
CAMDEN BAY
BEAUFORT SEA
BEAUFORT LAGOON
DEMARCATION BAY
KATAKTURUK R.
MARSH CREEK
CARTER CR.
HULAHULA RIVER
OKPILAK RIVER
JAGO RIVER
AICHILIK RIVER
EGAKSRAK RIVER
KONGAKUT RIVER
SADLEROCHIT MTNS.
SCHRADER LAKE
PETERS LK.
MT. CHAMBERLIN
MT. MICHELSON
MT. ISTO
ROMANZOF MTNS.
CANNING RIVER
MARSH FORK
RED SHEEP CR.
CANE CR.
TH MTNS.
CUPINE KE
ALASKA
CANADA
FIRTH RIVER
OLD CROW R.
NITSII DDHAA MTN.
UNJIK RIVER
CHANDALAR R.
ARCTIC VILLAGE
OLD JOHN LAKE
EAST FORK
SHIINJIK RIVER
COLEEN RIVER
UNITED STATES
CANADA
PORCUPINE RIVER
OLD RAMPART
CHRISTIAN RIVER
VENETIE

Foreword

IVISHAK, OKPILAK, AICHILIK, KONGAKUT. These rivers have kept their Native names, and for me they have magic. Along these and six or seven more, Debbie and Dennis Miller through the years have hiked and climbed and kayaked over a thousand miles. They have stood on top of Mount Michelson and seen, as Debbie said in a letter to me, "a spectacular view of mountains, coastal plain and Beaufort Sea, all in one far reaching glance."

Surely Debbie Miller is qualified to give us the story of the Arctic National Wildlife Refuge.

On one of their first treks she and Dennis hiked for 150 miles—studying the plain (20 to 40 miles wide), the permafrost, the rivers, the foothills, the nesting habitats of millions of birds, the glaciers, the boreal forest—and the Brooks Range which is the backdrop to a scene, an entity, a complete ecosystem, almost surely the only unit in North America of such matchless perfection.

Here the Native people, Indians and Eskimos, have lived for thousands of years without damaging their homeland. The over one hundred species of birds are still here—the polar bear, the grizzly, the wolverine, the wolf, and all the others. But the white man does not go gently into any place, and there is reason for alarm about the future.

Debbie will describe for you all the life of this rich corner of our arctic world.

But first let me quote from a speech Morris Udall gave in Arizona after his first trip to the Arctic.

"Not in our generation, not ever again, will we have a land and wildlife opportunity approaching the scope and importance of this one."

There, in those arctic valleys, there is room for pure unadulterated adventure and learning, for present and future generations. That is one reason for protecting the refuge. But more important, to my mind, would be our having courage enough, in the face of all challenges, to protect this region for the sake of *the land itself*, and the wildlife it supports.

I salute Debbie and Dennis, for all their travel and dedication, and Robin, who is almost surely the only one-year-old child in all history to talk to a wolf in its own world.

Will our society be wise enough to keep some of "The Great Country" empty of technology and full of life?

—Margaret E. Murie

Acknowledgments

I FEEL MOST FORTUNATE TO HAVE SPENT a substantial portion of the last thirteen years exploring the Arctic National Wildlife Refuge. First, and foremost, I'm deeply grateful for the numerous wilderness experiences, wildlife spectacles, and other natural gifts that are embraced within the boundaries of America's northernmost sanctuary. Without such precious time spent in the Arctic Refuge, this book would never have been written.

There are many individuals who contributed their time to the long book-writing process. As early as 1980, Dean Gottehrer, an inspiring journalism professor at the University of Alaska, encouraged me to write a book about the Arctic Refuge. I sincerely thank him for his initial support for this project and sage advice over the years.

Several individuals, including scientists in their respective fields and interested friends, reviewed particular chapters of the manuscript and offered helpful suggestions. Many thanks to Ken Whitten, Dave Klein, Ray Cameron, and Fran Mauer who contributed interesting information regarding caribou and other arctic species, and who critiqued the manuscript. Thanks to geologist Wes Wallace who enlightened me about the geology of the eastern Brooks Range and took the time to review what I had written about rocks. Philip Martin contributed information about birds who inhabit the coastal region of the Arctic Refuge and his comments were appreciated.

I'm also grateful to those who offered historical information about the Arctic Refuge and/or reviewed portions of the manuscript: Margaret Murie, Ave Thayer, George Collins, Ginny Wood, Celia Hunter, Dave Spencer, David Brower, and Jim Rhode. Their insight and personal involvement, concerning the movement to establish an arctic preserve, added much color to the past.

A number of other friends took interest in the manuscript and offered to review various chapters. Their suggestions and support during the project were much appreciated. Special thanks to Duncan Wanamaker, Doug Best, Sidney Stephens,

Sue Beck, and Nancy Witte. Thanks also to Richard Nelson who has written a number of books and whose guidance was much appreciated, and to Edgar Wayburn for taking a sincere interest in the manuscript.

Many interviews were conducted with scientists affiliated with the United States Fish and Wildlife Service, the Alaska Department of Fish and Game, the University of Alaska, and the Alaska Department of Environmental Conservation. The information distilled from numerous pages of interview notes clearly enriched the text. I'm grateful to Robert White, Steve Fancy, Dale Guthrie, Bob Weeden, Doug Schammel, Dave and Barbara Murray, Russ Oates, Alan Brackney, Greg Weiler, Steve Amstrup, Elaine Snyder-Conn, Pam Miller, John Hechtel, Wayne Heimer, Sverre Pedersen, and Brad Fristoe. A number of other staff members from these various agencies provided assistance during the course of my research. Thanks to all.

Margaret Murie deserves special recognition. Not only can we all be grateful for her past dedicated work to establish the original Arctic National Wildlife Range but I'm most thankful for her support and inspiration during the book-writing process. There were many times when her wonderful letters of encouragement would arrive in the mail, lifting my spirits and level of mental energy, and giving me the extra boost to finish what seemed like the never-ending manuscript.

A special thanks to my good friend Victoria Hand who created the maps for the book, and supported the writing project through her genuine interest and long-time friendship.

I'm also grateful to Danny Moses, my editor at Sierra Club Books for the first edition of *Midnight Wilderness*, who offered encouragement along with positive criticism as he reviewed chapter by chapter. His patience and easy-going nature were much appreciated, particularly during the final stages of the book when I was pregnant with our second child.

Finally, thanks to my husband, Dennis, who supported the family during the months of book writing, and to our two-year-old daughter Robin, who took interest in the book as she curiously thumbed, and sometimes scribbled, through pages of draft printouts while patiently waiting for Mom to finish just one more paragraph. Then, the book might not have been completed without her faithful two-hour naps.

March 1989

Many thanks to Helen Cherullo whose vision and dedication as a publisher, and whose love for the protection of wild places, is an inspiration to all of us. A heartfelt thanks to her, Kate Rogers, and the great editorial staff at The Mountaineers

Books for creating a new twentieth anniversary edition of *Midnight Wilderness*.

I'm also grateful to those who have raised their voices across America, expressing their commitment to protect the stunning wilderness and wildlife values of the Arctic National Wildlife Refuge. May this book continue to serve its purpose, introducing new readers to the wonders of this extraordinary place.

A huge thanks to the unending, dedicated work of the Alaska Wilderness League, the Northern Alaska Environmental Center, and other conservation groups whose efforts have successfully kept the Arctic Refuge wild and free of industrialization. It is hoped that this book will continue to be a useful tool for wilderness advocates and educators.

Another thanks to my husband, Dennis, and our daughters, Robin and Casey, for sharing all the memorable experiences and adventures in the Arctic Refuge as a family. It is my hope that our daughters' children and grandchildren will have the opportunity to visit the Arctic Refuge in its truly wild and natural state.

Last, a special, unending thanks to the late Margaret E. Murie for her work and dedication in establishing the original Arctic National Wildlife Range with her husband, Olaus Murie. Their work will always be remembered, and future generations will carry on the torch to protect this magnificent area in their honor.

March 2011

Introduction

I REMEMBER THOSE KEY DECISIONS in past years that have marked a turning point in my life. One of the most important decisions, which profoundly changed my view of the natural world and the quality of my lifestyle, was the decision to move to and ultimately settle in Alaska.

In the summer of 1975, my husband, Dennis, and I resigned from our teaching positions at Marin Country Day School in Corte Madera, California, and headed to Alaska after having read stacks of books on the Far North, we were convinced that the only way to fully experience Alaska and its great wilderness was to live there. We had concluded that the intriguing forty-ninth state was too far away, too extraordinary a place to merely visit on a summer vacation. So we packed up our camper with the essentials, stored the boxes of unnecessary wedding gifts, said our goodbyes to relatives and friends, and gradually left the California freeways and the pace of congested cities in the rearview mirror.

Shortly after arriving in Alaska, we learned that one of the state's most isolated Athabaskan Indian villages, Arctic Village, was in need of an elementary-school teacher. We pulled out our map and discovered that this small village of 125 residents was located in the remote northeastern corner of Alaska, on the south slope of the Brooks Range, just beyond the vast Arctic National Wildlife Refuge. We agreed that Arctic Village was beautifully situated in one of the wildest regions in North America. I couldn't consider passing up the teaching opportunity, particularly when we learned that Arctic Village was a friendly community whose residents had preserved their traditional subsistence-based culture and native language.

Within a few weeks of our arrival, we purchased a year's supply of food, bought clothing for sub-zero weather, and headed north in a small four-seater bush plane. I can still vividly remember the excitement of that first flight: viewing the endless stretch of wild country; spotting our first caribou and Dall sheep; following the exquisite Chandalar River as it sparkled in autumn color; soaking up our first view of the Brooks Range with its forever span of peaks and valleys; and landing on the gravel airstrip where a truckload of curious, dark-eyed children waited to size up the new teacher.

We cherish our years spent in Arctic Village and are grateful for the rich experiences shared with a remarkable group of people whose ancestors have survived in this remote region for thousands of years. The people of Arctic Village taught us much in the way of hunting and survival skills, skin tanning and sewing, and other fascinating aspects of their subsistence-oriented culture. I had initially intended to devote a section of this book to those native people, both Athabaskan Indians and the more northerly Inupiat Eskimos, who are intimately bonded to their homeland, which includes areas within the Arctic National Wildlife Refuge. After much consideration, however, I decided that the subject, given proper coverage, could certainly fill the pages of yet another book. Throughout this book I have sprinkled pieces of information about the historical use of the Arctic Refuge by native people, but the reader should realize it is merely a sprinkling of two very rich and complex cultures.

Living in and near Arctic Village gave us a tremendous opportunity to explore the 19-million-acre Arctic National Wildlife Refuge, from its northern coastal plain, which sweeps to the ice-mantled Beaufort Sea, to its awesome glaciated peaks within the Romanzof Mountains, to its boreal forests that ring the south slopes of the Brooks Range. Through the course of the last thirteen years we have spent a portion of every summer traveling by foot or kayak through the refuge, crossing hundreds of miles of the purest and largest wilderness area remaining in the United States.

During our many arctic journeys within the refuge, we rarely came into contact with other human travelers, or even old footprints; yet, over time, we have witnessed tens of thousands of caribou walking by our tent; seen scores of Dall sheep, grizzly bears, and the occasional passing wolf; and watched many of the hundreds of thousands of migratory birds hatch their young. One could spend a lifetime exploring the nameless river valleys and peaks of the refuge and not see it all. The Arctic National Wildlife Refuge is the essence of America's wilderness heritage. It is a rare sanctuary that I have had the privilege to experience and have grown to love deeply.

In recent years the coastal plain of the refuge has been threatened with the increasing possibility of oil exploration and development. Such industrialization on America's only virgin stretch of arctic coastline would irreversibly destroy the wilderness character of the area, and undoubtedly disturb, displace, or reduce wildlife populations that utilize this most biologically productive zone of the refuge. At the time of the writing of this book, it is uncertain what the fate of the refuge will be, as the development question lies in the hands of Congress.

Beginning with the coastal plain zone, this book will take you on a series of journeys through the refuge, in Part I, you'll walk across the coastal plain, plod through the tussocks, and experience this region as it teems with tens of thousands of newborn caribou calves and other wildlife. In Part II, you'll travel from the coastal plain into the Romanzof Mountains, trek through pristine valleys, and cross America's northernmost glaciers. You'll follow rivers such as the Okpilak, Hulahula, and Chandalar to their headwaters, and cross the rugged spine of the Brooks Range. Part III focuses on the history of the fight to establish the original Arctic National Wildlife Range back in the late 1950s, and includes the account of a twenty-one-year-old mystery that was solved during one of our backpacking trips along the Ivishak River. Part IV focuses on a week of living within the neighboring Prudhoe Bay oil fields in contrast to dwelling within the untouched wilderness of the refuge as well as seeing this rare sanctuary through the eyes of our young daughter, Robin.

Most of the chapters are based on years of journal writings, enriched with natural history information. It is hoped that those reading this book will gain a deeper appreciation for this unique arctic region and its well-adapted inhabitants. If you never have the opportunity to visit the refuge, it is hoped that you will vicariously experience the arctic, and enjoy discovering the special wilderness and wildlife values of the area through these personal accounts. For you who may tuck this book in your backpack on a trip through the refuge, it is hoped that you gain a greater understanding for this magnificent region, and through your own experiences, share with me the great sense of discovery and adventure that the refuge possesses.

Finally, it is the author's personal goal that by reading this book individuals will become more fully aware of the increasing importance of protecting the extraordinary wilderness, wildlife, and recreational values of this world-class arctic refuge for future generations of wildlife and man. It is one of the few remaining wild regions on this planet and should be preserved in its whole and natural state.

—Debbie Miller, 1989

The year 2000 marks the twenty-fifth anniversary of our first journey through the Arctic Refuge. Our family continues to visit this extraordinary area, and each trip holds special memories. Our two daughters, now fourteen and eleven, have witnessed the migration of thousands of caribou, watched wolves and grizzlies, fished for arctic grayling, and scaled peaks with Dall sheep grazing nearby. Our daughter Robin asks, "Why would anyone ever think of developing this place?"

The year 2000 also marks the anniversary of two monumental achievements: the twentieth anniversary of the passage of the Alaska National Interest Lands Conservation Act, and the fortieth anniversary of the establishment of the original Arctic National Wildlife Range. These two acts set aside the greatest wilderness areas that exist on the planet. More than 100 million acres of national parks, refuges, and forests were designated under the 1980 Alaska Lands Act, including the Arctic Refuge, which doubled the size of the original Arctic Range. The Alaska Lands Act is the greatest lands conservation act ever enacted in the history of mankind—a great cause for celebration.

One summer, former President Jimmy Carter and his wife, Rosalynn, visited our family in the Arctic Refuge. It was under President Carter's leadership that the Alaska Lands Act was enacted, along with the dedication of countless conservationists who fought diligently to protect the wilderness and wildlife values of Alaska. While visiting the Arctic Refuge, the Carters had the incredible opportunity to witness the gathering of the Porcupine caribou herd. As they watched thousands of animals flood the land, President Carter was awestruck by the wildlife spectacle: "The closest thing I've seen to this is Africa's Serengeti Plain. Oil development can never be allowed here."

Over the past decade there have been many political battles concerning proposed oil drilling in the Arctic Refuge. Development forces have pushed steadily to open the Arctic Refuge coastal plain to oil drilling, while conservation groups and the Athabaskan Gwich'in people of northeast Alaska and Canada have worked tirelessly to protect the coastal plain as wilderness. Support for protecting the Arctic Refuge is at an all-time high, yet as in 1990, the final decision rests in the hands of Congress and the President.

While oil development has increased in the Arctic, the Arctic Refuge remains unchanged. The wilderness is still vast, the beauty sheer. The caribou still give birth to thousands of calves each year, the wildlife spectacles continue.

Never before has there been a better time to permanently protect the Arctic Refuge coastal plain, an extraordinary land worth saving for the world.

—Debbie Miller, 2000

Part One

❖

NORTHERN BIRTHPLACE: THE COASTAL PLAIN

Pacific loon's nest

1

Migrations Flood the Land

WE WALK ACROSS A SEA OF TUNDRA, stunned by the open, boundless vista and never-setting sun. Thousands of square miles of wilderness surround us on this vast coastal plain of the Arctic National Wildlife Refuge. Like the desert, the unforested tundra possesses the beauty of a simple, uncomplicated landscape where the eye can embrace the surroundings, absent of any obstructions, and the mind can drift with the flow of the expansive landform. The uncluttered freedom of space, unlimited visibility, and naked, yet living, tundra are awesome.

Dennis and I had left the grind of civilization back in Fairbanks, about four hundred miles to the south. Shortly after flying north of the city, we crossed the last paved road and a few remote cabins; then we continued on across an endless stretch of wild: over the White Mountains; across the Yukon Flats, with its myriad of lakes, ponds, and sloughs; above the invisible Arctic Circle; past the northern limit of spruce trees; over the serrated crest of the Brooks Range; down its glacially carved valleys; and across the expansive coastal plain, which sweeps from the foothills to the Beaufort Sea. There was only a hint of human presence along our flight route. Across the northeast part of Alaska, the region is punctuated by only a few small Athabaskan Indian villages: Birch Creek, Fort Yukon, Venetie, Arctic Village, and, finally, the Inupiat Eskimo village of Kaktovik on the shore of the Beaufort Sea.

On the coastal plain there is an overwhelming sense that we have been thrown back to a more primitive age: to an age when man roamed the earth in small numbers, faced other predators, and survived in a land where nature was the governing force; an age when man represented only a small fraction of global life. Here, man has neither multiplied, manipulated, conquered, nor consumed this most northern terrestrial ecosystem. Here, other animal species outnumber man, from the migrating caribou to the ubiquitous mosquito.

If we travel a dozen miles to the north we reach land's end, on the shore of the ice-choked Beaufort Sea, the edge of the North American continent. Beyond this edge we enter the world of seals and polar bears, of bowhead and beluga whales, and of an icescape deeply understood by the Inupiat Eskimos who have survived in this region for centuries.

Gazing beyond the open tundra plain, I stare at the shimmering endless pack ice, curving toward the North Pole, some fourteen hundred miles away. Beneath a dome of larkspur-blue sky, on this final fringe of tundra, with our civilized world far behind us, a tremendous feeling of wildness surges through me. We are truly on the edge of the most remote wilderness area remaining in the United States. We are on top of the world.

Behind us, some twenty-five miles to the southwest, beyond the foothills, glaciated peaks of the Brooks Range dramatically rise to heights exceeding 9,000 feet, forming an east–west rampart along this narrow swath of coastal plain. Because we stand near sea level, the mountain relief seems higher than what the map tells us; it is much like viewing the Cascade Range from Puget Sound. The mountains are not in the 20,000 foot league like the Himalayas, yet they strike you as formidable, luring. These mountains, mantled with small but active glaciers, are the highest peaks within the entire Brooks Range.

Only a few months ago this region was locked in winter, with temperatures of fifty and sixty degrees below zero, and the wind chill factor dropping those figures to as low as one hundred below zero. Even as late as May the mean temperature in Kaktovik, thirty-five miles to the northwest, is merely twenty degrees. Only the most well-adapted creatures, like the muskox, polar bear, arctic fox, and Inupiat Eskimo, can survive in the high Arctic as year-round residents. The oppressive arctic winter lingers from mid-September through May, making the coastal plain one of the coldest, and most lethal, places on the planet to live.

On this July afternoon, with temperatures soaring into the 70s, that frigid world seems light-years away. In the dead of winter, this land's white crust of snow and ice is void of direct sunlight. For almost two months a muskox sees only its moonlit shadow. Yet now this polar world tilts toward eternal sunshine, and days are timeless. The snow has vanished, brown stubble has been overtaken by fresh green growth, and small, brightly colored wildflowers nestle their faces among the mosses, lichens, and dwarfed shrubs that crawl along the tundra. The songs of nesting birds greet us from every direction—cheery Lapland longspurs, peeping sandpipers, and the soft-spoken savannah sparrows.

The arctic winter releases its icy grip for only a few brief summer months. In June the land responds to perpetual daylight, offering an explosion of plant life and insects for hundreds of thousands of migratory birds who have traveled here from places as far away as South America and Asia. The coastal plain is also an ancient birthplace for an estimated 180,000 migratory caribou of the Porcupine herd, a herd that knows no boundaries, wandering thousands of miles each year within Alaska and Canada's Yukon and Northwest Territories.

The air is clouded with countless, humming, blood-thirsty mosquitoes. Walking along the banks of the Egaksrak River, we long for a breeze to blow away the aggravating bugs. I purposely take shallow breaths. Even though I have put on insect repellent, deep breaths bring the mosquitoes fluttering down my throat. I tried a headnet once but quickly decided that I didn't like viewing this world through army-green mesh. Why look at the wildest country in America through a screen door?

Although humans, caribou, and other species are bothered by mosquitoes, they are an extremely important food source for many of the 108 species of birds who migrate to the coastal plain each year. Mosquitoes are to birds what krill is to whales. If you remind yourself of their importance within the food chain, you may grow to appreciate and tolerate their existence, although it's unlikely that you'll swat at them any less.

While gagging on a couple, I hear a faint peeping sound within arm's reach. I crouch down low to the tundra, careful not to make any sudden movements. There between the tussock mounds is a semipalmated sandpiper chick wobbling through the grasses, pecking at mosquitoes. The chick can be only a few days old, yet it instinctively knows how to spike insects. Its beak automatically jabs at any mosquito pausing on a blade of grass, as if the motion is involuntary.

I watch this buff-colored ball of down teeter across the uneven landscape, occasionally losing its balance like a staggering drunk. Somewhere in the tussocks is a camouflaged nest crafted by the chick's parents. I study the grasses closely, trying to locate the invisible nest. Then the mother of the chick lands on a nearby tussock on her partially webbed feet. With alarm she peeps until I slowly move away from her chick.

When grown, these small sandpipers, about five to six inches long, travel thousands of miles each year from places like the coast of Venezuela to breed in the Arctic. It is impossible to imagine that this tiny fluff of life, so vulnerable on the open tundra, will soon be able to spread its wings and begin its southern migration to unknown points along the southeast coast of the United States, the Caribbean Islands, and as far as South America. The chick is just one of hundreds of thousands of birds scattered across the tundra who are to fledge and begin their long journeys down river valleys, through mountain ranges, across basins and prairies, and along the scribble of coastlines. Each delicate ball of down is a miracle.

This particular pair of sandpipers built their nest and timed the hatching of their young to coincide perfectly with the emergence of mosquitoes and midges.

The chicks rely solely on the insects as their source of food. If they hatch too early, before the bugs have emerged, or during an arctic storm when they can't venture out to find insects, they die. Like other shorebirds, these sandpiper chicks are independent feeders from the moment they hatch.

We pause on the banks of the braided Egaksrak River, listening to the water murmur as it flows past rounded stones of mountain origin that have rolled out onto the plain. Several weeks ago the river was raging with spring snowmelt, boulders rumbling against one another in the surge. Today the river runs calmly around gravel bars no longer swallowed by torrents. Clumps of sticks and grasses are matted in willow bushes along the bars, revealing the high-water line of the swollen river, about three or four feet above the current level.

As I look east across the sweep of tundra toward the Canadian border, about forty miles away, I see the entire horizon waver through heat waves as if we are on a desert plain. At first glance the liquid figures, bubbling up from the tundra, appear to be only part of a huge mirage. Looking closer, I realize that this is more than an illusion. It is a living, breathing, pulsating mass of animals. Through the heat-waves, thousands of silhouetted caribou are melting into the sky.

My heart races. A large segment of the Porcupine caribou herd is moving toward us. Although we have spent portions of many years within the Arctic Refuge, we've never been lucky enough to observe large numbers of aggregating caribou from the ground. It is the chance of a lifetime. I fumble for my camera, but I'm too over-whelmed to think about light meters and viewfinders. Caribou by the thousands are coming toward us. The spectacle fills my mind, as does the landscape in front of me. Forget the pictures.

Shoulder to shoulder, they stream across the tundra, a tide of undulating bodies, gracefully moving across miles of tussock-covered terrain. A whispering chant of grunts and bleats fills the summer air in rhythm with the herd's steady pace. As they come closer the voices grow louder. It is difficult to judge distances on this expansive coastal plain, but they appear to be less than one mile away. My heart continues to race as I stand spellbound. They are coming closer; the sounds grow louder. One of the world's greatest wildlife spectacles. And we are here to witness it. The minutes seem like hours.

Soon the tundra before us is flooded with caribou. There is barely a trace of ground through the mass and clutter of hooves. Where the tundra appeared empty only a few minutes ago, now it is a moving mass of flesh and blood. Month-old lanky calves bleat for their mothers. Mothers extend their necks and muzzles, mooing for their calves. Most of the caribou are cows with calves, although we spot a number of yearlings, juveniles, and bulls mixed in with the herd.

As the caribou approach, the vocal sounds of the animals are as overwhelming as the scene itself. In addition to the cattle-like bleating and mooing, there is a tremendous amount of grunting, snorting, bellowing, and even coughing. The sounds are reminiscent of the time when, as a teenager, I had the exciting opportunity to assist with a cattle roundup on a Montana ranch. I still remember the din of the cattle's voices as we funneled the animals into a corral. Yet the sounds of the caribou are more varied.

Perhaps the most unique sound is the strange clicking of hooves. No other member of the deer family produces this kind of clicking, which resembles the sound of castanets or of drumsticks beating one another. The sound is caused by a springing ligament, characteristic in ungulates, that slips over bones within the hoof when the caribou walks. The clicking of tens of thousands of hooves blending with the variety of vocalizations, in rhythm with the herd's gait, creates a caribou symphony on the tundra.

The voices and clicking hooves crescendo as the large aggregation of caribou literally envelops us near the river. Soon we are an island in a sea of swarming animals. We would later find out that ten to twenty thousand animals were encircling us. They are everywhere. Immersed within the herd, we are close enough to hear small calves panting from the heat as they trot past, caribou teeth clipping and chewing tundra plants, and tussock grasses being compressed beneath the hooves.

We watch caribou quiver and shake off the harassing mosquitoes, like animals shaking water out of wet fur. Some calves attempt to suckle their mothers for a few moments, but the cows move on with the calves scurrying behind. Stately bulls with new sets of velvet antlers pause and look curiously at us, then stride on.

The caribou seem to be eating on the run, not because of our presence, but because of the mosquitoes. They stop to forage for a few moments, shaking their bodies while they eat, then move on. They are angling toward the coast, where it is breezier, cooler. There they will find some relief from the plague of bugs. Caribou biologists label cool, breezy coastline areas as insect-relief habitat, a sort of war buffer zone for the caribou. Under severe mosquito harassment, caribou can lose up to a quart of blood per week, an extra-heavy burden for the nursing cows, who are recovering from pregnancy and delivery. The loss of blood can also cause slow growth in calves.

As countless caribou pass, most appear to be oblivious to our presence, but some pause momentarily to look at us. Experiencing eye contact with any wild animal in an unprovoked situation can be a treasurable moment. Any perceived boundaries between man and other members of the animal kingdom are temporarily suspended, if only for an instant. During that instant, there is an intense desire to communicate with the other being, coupled with a longing to understand what dwells in the mind of that other creature, be it a squirrel in your backyard or a gray whale at sea. In that same instant there is a tremendous urge to somehow bridge the gap between you and the other creature, to erase all the eons of evolutionary changes, to enter the realm of the other creature's world in a non-threatening way, and to somehow understand one another.

Looking into the large brown eyes of caribou calves as they pause and curiously look at me, I stand calmly, hoping that our eyes will meet for just a little longer. If only I could convey my feelings of emotion and respect for these remarkable creatures. It is believed that the ancestors of these calves have delivered their young on this virgin stretch of coastal plain for at least 2 million years. These tundra dwellers are one of the few large mammals that have established year-round populations in the Arctic. The Porcupine caribou herd is one of the largest herds in Alaska and Canada that are often referred to as barren-ground caribou.

Where and how the caribou have evolved to live over the centuries is extraordinary. They have adapted to some of the harshest climatic conditions in the world, traveling great distances to seek forage, rigorously cratering through snow most of the year to eat lichens, their most important winter food. In the springtime the caribou wintering grounds look like a moonscape with countless craters pocking the snowfields

U.S. Fish and Wildlife Service (USFWS) studies have documented, through satellite radio collars, that a caribou from the Porcupine herd travels on the average about twenty-seven hundred miles per year. No other terrestrial mammal in the world travels so extensively in the course of a year. The caribou's closest rival is the wildebeest in Africa. Studies have also documented that the light-bodied, long-legged caribou are the most efficient walkers of all land mammals on earth. Like a well-conditioned marathon runner, caribou travel far by burning up the required energy in the most cost-efficient manner.

Biologists believe that the Porcupine herd has chosen the 2-million-acre coastal plain of the Arctic Refuge as its preferred birthplace for several reasons. The coastal plain has a scarcity of predators when compared to adjacent mountainous areas, where wolves and grizzly bears make their homes, and where golden eagles use rising air currents along mountain edges to effortlessly soar above their prey. Studies suggest that the caribou's mortality rate is significantly higher in the foothills and mountains of the Brooks Range than it is on the coastal plain.

Also, when the insects are at their worst, the coastal plain offers good insect-relief habitat on the beaches and inland along numerous river gravel bars. Caribou

Caribou on coastal plain

have optimal access to cooler, breezier habitat, and can escape the harassing clouds of mosquitoes.

As caribou arrive on the coastal plain calving grounds in late May, the tundra frequently has patchy snow conditions. Such a setting offers camouflage for the caribou. Major predators like the grizzly bear, wolf, and golden eagle have a more difficult time spotting caribou in the mottled setting. Plus, the caribou are able to take advantage of foraging on the first green plants that emerge from the thawing plain, coinciding with the plant's most nutritious, budding state.

Robert White and David Klein, at the University of Alaska, have extensively studied the nutritional needs of caribou and other well-adapted mammals in the northern latitudes. They point out that the coastal plain area offers an excellent variety of high-quality forage that is particularly important for nursing cows, for milk production, and for calf growth. After a winter diet, primarily of lichens, the coastal plain offers a salad bar for the caribou. Caribou eat about three times as much forage during the brief summer green-up as they consume during the winter months.

During June and July, a progression of desirable plants high in protein, carbohydrates, and vitamins emerges from the snow-free coastal plain, beginning in the foothills and extending north to the Beaufort Sea. As snow first begins to melt at the higher elevations, millions of tussock heads are exposed. The bumpy tussocks, generally one to two feet high, look like miniature conical thatched huts. They dominate the coastal plain the way dense barnacles cover a seashore rock.

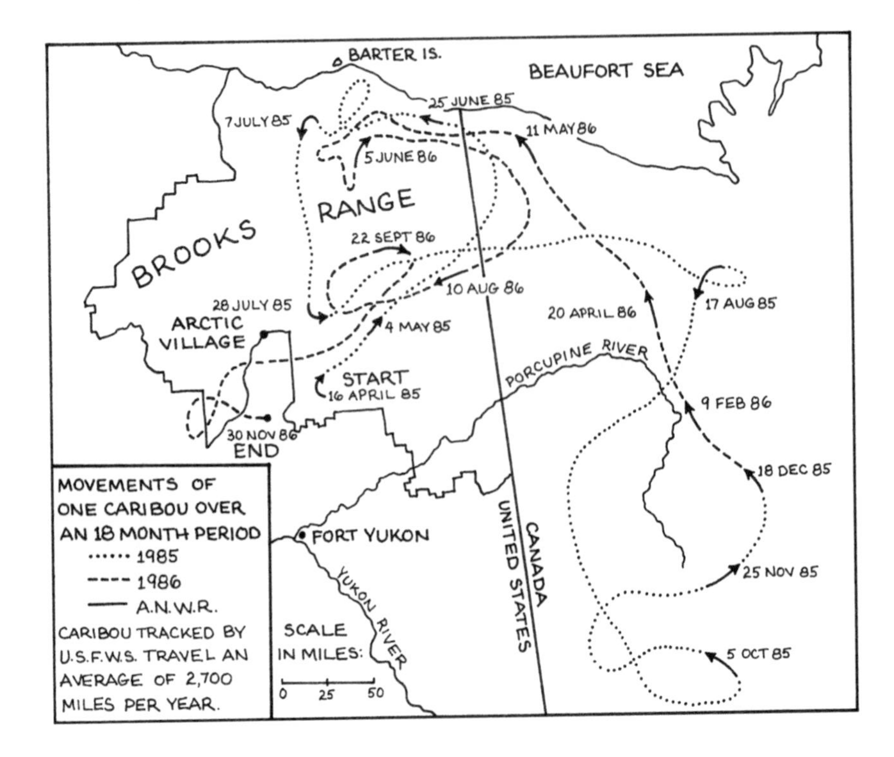

Each tussock offers a microcosm of plant life, for foraging caribou, in particular, the prolific cotton grass, which is a high-protein source. The dark-colored floral buds of the cotton grass are designed to absorb heat from the sun when air temperatures may be hovering around freezing. The elevated tussock, a product of the cotton grass, offers a natural greenhouse effect to other plants that are rooted within the mounds. Plants that have chosen tussocks for their homes are able to maximize the sun's energy. Their feet warm up sooner, and, therefore, they can bud earlier. When I trowel the rows and hills within my vegetable garden at home, I use the tussock concept to fashion my soil.

Low-growing willows are another important food source. The coastal plain within the Arctic Refuge has a higher proportion of rivers and creeks with a higher density of riparian willow stands than the rest of Alaska's North Slope. Willow leaves are higher in protein than grasses or sedges, and high in vitamins A and C.

The caribou's menu also includes such foods as low-bush cranberry leaves, lupine buds, and the floral parts of dryas. Lousewort, or bumblebee flower, is one of the first flowering plants to emerge from the tundra, often when snow is still on the ground. Dr. Klein refers to the lousewort as "an ice cream food" for the caribou

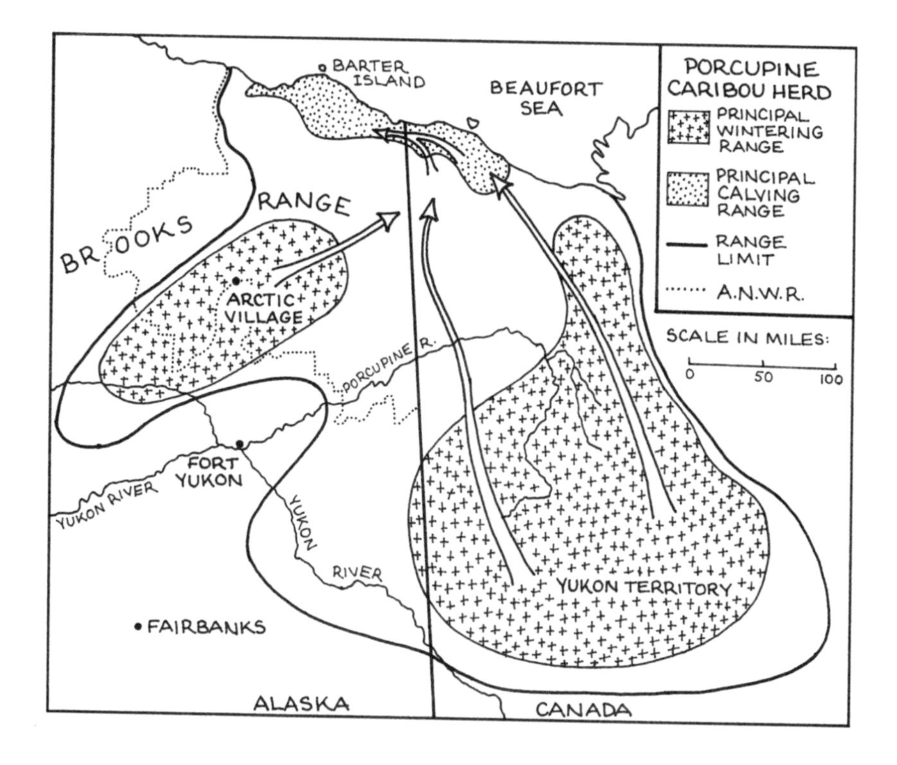

because it is one of the best sources of both protein and carbohydrates and is easily digested. It is not a high-density plant on the tundra, but if a caribou sees one, it will go out of its way to nip the plant.

Watching the calves trot by me on their long bar-stool legs, I'm amazed that only four weeks ago they came out of the womb, slipping onto a patch of tundra, weighing about fifteen pounds. Because of the communal nature of caribou herds, caribou breed in a very synchronized manner each fall, and pregnant cows have evolved so that they all deliver their calves at about the same time. The calves are usually born within a ten-day to two-week period during late May and early June, after the pregnant cows have completed their arduous spring migration from their winter range.

Dennis, a wildlife survey pilot, has often marveled at the calving activities. He describes that one day in June the coastal plain will be filled with tens of thousands of pregnant cows, with very few calves in sight, while the next day the land will be flooded with thousands of newborn calves. Indeed, biologists calculate that the majority of the yearly crop of forty to fifty thousand calves are born over a two- to three-day period known as the peak of calving.

Calving in such large numbers gives caribou an advantage over their predators. Biologists refer to this defense technique as "swamping" the predators. With tens of thousands of calves and mothers in one general area, a cow/calf pair within the masses has a better chance of not getting caught, like children playing the game of "sharks and minnows." Also by grouping together in one spot, the caribou expose themselves to a minimum number of predators.

Most cows deliver their calves while lying down, although some will drop them on the tundra from a standing position. Cows are in labor for a relatively brief period of time, usually about an hour. Unlike moose and other members of the deer family, who often carry twins, the caribou have evolved to produce only one calf.

Caribou are thought to produce a single calf for several reasons. Since caribou congregate in large numbers, it would be difficult for a cow to keep track of more than one calf; and a lost calf becomes more susceptible to predatory attacks and can live for only two to three days without its mother. Another theory suggests that at the time of calving, pregnant cows have an extremely limited amount of fat reserves as a result of the long drain of winter and having just completed their spring migration. Such energy demands may make it physically impossible for the cows to successfully carry and nurse more than one calf.

By the time the pregnant cows reach their calving grounds, they have plodded through snow for eight to nine months, and traveled thousands of miles through their winter range, which extends south of the Brooks Range and into the Ogilvie Mountains in Canada's central Yukon Territory. In George Calef's eloquent book *Caribou and the Barren-Lands*, he follows the 1980 spring migration of the Porcupine herd. He notes that some pregnant cows were so exhausted from migrating through deep snow that he could walk up and touch them because they slept so soundly.

Within a few hours after birth, caribou calves are following their mothers, suckling them frequently. Caribou milk is high in protein and fat—two to three times as rich as dairy-cow milk. The tawny calves with their stiltlike legs, suckle for two to three minutes at a time, several times an hour. This contrasts with members of the deer family who live in forested or brushy areas where more dense forage is available.

Some deer, such as elk and black-tailed deer, hide their young in thickets to escape predators. The mothers forage nearby and periodically return to nurse their calves for longer periods than caribou do, perhaps only four times a day. Barren-ground caribou can't afford such luxury. They move great distances each day to obtain enough forage and to rid themselves of harassing insects. Caribou will hide their newborns among the tussocks until they are about two days old, but after that, there is no place on the open tundra for the mothers to stash their fast-growing calves. To survive, calves must follow their mothers at all times.

Cows, calves, yearlings, and, occasionally, bulls continue to stream past us, pouring over the Egaksrak River bank, splashing into the water. I think that this must be what Africa is like. For a moment my mind drifts to the Pleistocene era, when woolly mammoths, camels, ponies, saber-toothed cats, and lions roamed across the glacier-free coastal plain and interior Alaska. I picture a saber-toothed cat, with its long, piercing canines, chasing caribou across the tundra; camels loping along a more arid plain; woolly mammoths the size of elephants, with yard-long hair, grazing near the windswept undulating sand dunes that once were a characteristic landform.

These animals, along with today's caribou, muskoxen, wolves, Dall sheep, and others, lived together in this region until the late Pleistocene. During the close of the most recent glacial period, about eleven to twelve thousand years ago, the last of the mammoths, camels, ponies, lions, and saber-toothed cats gradually became extinct. It is uncertain why there appears to be no trace of these animals in fossil records younger than eleven thousand years. One theory suggests that climate and vegetation changes played a key factor, while another theory proposes that hunting by aboriginal people, migrating into Alaska from Siberia, was a cause.

While teaching in Arctic Village on the southern boundary of the Arctic Refuge, I remember an Athabaskan resident showing me part of a mammoth tusk, about half as long as his arm and as thick as his fist, that he had discovered near a river cutbank. The well-preserved tusk had been embedded in frozen silt for centuries, although it had discolored to a muddy auburn brown. On a sunless December day, at forty-five degrees below zero, he joked about the idea of hunting elephants out his back door.

Back on the coastal plain, caribou swim single file across the river, sending up a glint of spray in the midday sun as they splash into a deep river channel. Although some caribou drown each year during river crossings, they are known to be excellent swimmers. Caribou have unusually broad hooves which help them float on snow, dig through snow for their winter forage, and swim across rivers. In the summer the hooves serve as paddles; in the winter they turn into shovels. Caribou also have hollow, coarse hairs that increase their buoyancy when they swim, and minimize their heat loss during winter, acting like air chambers within a good sleeping bag.

Out of the thousands I focus on one calf, which seems to be struggling to keep up with its mother. After paddling across the river, the calf ascends the bank and shakes the water out of its furry coat. Appearing lost and alarmed, it stands bleating on the edge of the bank. It listens to thousands of unrecognizable grunts and bleats, its ears flapping up and down, attempting to decipher the particular call of its mother. The calf anxiously trots up the riverbank, sniffing the ground occasionally, then bleats. How will this small calf find its mother in the mass of animals?

After several minutes, the calf swims back across the river, in the same channel, to continue the search. Swimming to a gravel bar, it finds another cow and calf and attempts to suckle the cow. The cow slowly moves away but does not reject the lost calf. The cow and her calf enter the river, and the lost calf follows. When they reach the opposite shore, the three run toward the bulk of the herd. At least for the moment it looks as though the lost calf has found a surrogate mother. Normally caribou mothers will reject a foreign calf, pushing it away.

Within a half hour most of the herd has passed us. A few crippled stragglers bring up the rear, limping over the tussocks. Left in the herd's wake is another lone calf, confused and bleating. From the retreating mass of caribou appears one cow, who finds the missing calf. After a short period of suckling, the reunited pair run off to catch up with the herd.

Gradually the swarm of animals leaves us, ebbing away from the river. Their voices become muffled, then faint. Their white rumps grow smaller and smaller, until they fade into a massive shadow across the expanse. For several minutes caribou fill the distant horizon, their bodies now silhouetted against the white, sparkling pack ice. Then they vanish as quickly as they had appeared.

We each remember certain dynamic experiences, extraordinary events, or rare moments that are imprinted within our brains like fossils within their rocks. In our mind's eye such profound experiences, when recalled, are vivid, timeless events that fade little through the years. Such memories often flood the mind and heart with overwhelming emotion.

Witnessing the postcalving aggregation of the Porcupine caribou herd is one of those truly amazing events that I will never forget. As the last of the herd disappears into the ice-mantled horizon, I stand frozen on the tundra, pondering the incredible spectacle. Tens of thousands of month-old calves, nursing mothers, stately bulls returning to their northern birthplace. They have journeyed here for centuries to bear their young. It is life in the Arctic at its fullest moment of creation, in the wildest of landscapes. I am so grateful for being here. Tears stream down my face from the beauty and magnitude of it all.

Yet, it is more than the magnificent spectacle that evokes my emotion. This ancient, northern birthplace is where the oil industry desires to drill. The thought of oil rigs, drilling pads, and a maze of pipelines and roads on this northernmost horizon, on America's only virgin stretch of arctic coastal plain, overwhelms me with sadness. It seems as sacrilegious as driving a motorcycle into the Sistine Chapel. It is as inconceivable as constructing an oil field within Yosemite Valley.

2

Beaufort Lagoon

LATER IN THE DAY WE RETURN to Beaufort Lagoon in our small plane, a two-seater Piper Super Cub. We follow the Egaksrak River to where it empties, with the neighboring Aichilik River and other smaller streams, into a lagoon. Numerous braided river channels snake through the shoals, forming an arctic delta. The burning glare off the water keeps me squinting as I watch for birds. We spot ducks, geese, a pair of tundra swans, and many shorebirds near the mouth of the river and in nearby tundra ponds.

Beaufort Lagoon is separated from the Beaufort Sea by narrow islands of coarse sand and gravel, about ten to twenty-five yards wide, that parallel the coast for several miles. Shorefast sea ice hems the barrier islands on the north side, with only a few open leads visible. Large mantles of sea ice, several feet thick, have gnawed at the exposed islands, and shoveled their leading edges beneath the sand and gravel. Within the half-mile-wide lagoon the ice is breaking up, and there is much open water. Some of the thin, melting floes are almost transparent.

As we fly along Beaufort Lagoon, I picture Sir John Franklin with his British expedition party in their two open boats dodging ice floes in this very lagoon, dragging their boats across shoals, and visiting Inupiat groups that were scattered along the coast in clusters of sod houses. Franklin was the first white explorer to visit this part of Alaska's arctic coast, during July and August of 1826. He named Beaufort Lagoon (Franklin referred to it as a bay) and Beaufort Sea after a friend and naval hydrographer, Captain Francis Beaufort. Traveling from Canada's McKenzie River delta, Franklin ventured west along the arctic coast to Point Beechey. Today the Prudhoe Bay oil fields envelop the Point Beechey area, about 175 miles west of Beaufort Lagoon.

We follow the lagoon's shoreline for a few miles west until we reach an abandoned Distant Early Warning (DEW) line site. This small military radar station, and a chain of other stations along the arctic coast, were built back in the early 1950s during the cold war era. A few metal, windowless buildings still remain and are occasionally used by the USFWS staff as a base for summer fieldwork. The buildings and scattered rusted military equipment at Beaufort Lagoon will eventually be removed by the Army Corps of Engineers as part of a cleanup project. The site is an eyesore on the open tundra.

We land on a short gravel airstrip next to the DEW station, on a small point jutting into the lagoon. For now, Dennis is based at the lagoon with several USFWS and Alaska Department of Fish and Game (ADF&G) biologists. They are here to conduct an aerial census of the Porcupine caribou herd and to study the movements of more than one hundred radio-collared animals. I am here as a volunteer to assist with the census work, and to visit Dennis, who spends a large portion of each summer flying wildlife surveys within the refuge.

These biological studies are a result of a congressional mandate under Section 1002 of the 1986 Alaska National Interest Lands Conservation Act (ANILCA), directing that 1.5 million acres of the Arctic Refuge coastal plain be assessed for its biological importance and potential oil and gas reserves. This same landmark act added 97 million acres to Alaska's National Park and National Wildlife Refuge systems. ANILCA more than doubled the size of the original Arctic National Wildlife Range, to about 19 million acres, reclassified the range as a refuge, and designated 8 million of those acres as wilderness.

However, as part of the 1980 legislative compromise between oil industry and environmental factions, 1.5 million acres of the arctic coastal plain were put in an "undecided" category. Congress authorized the Department of Interior to study, assess, and ultimately recommend whether the "1002 area" (referred to as ten-oh-two) should be opened for oil and gas leasing, left as it is now—de facto wilderness—or legally protected as wilderness under the National Wilderness Preservation System. In 1983, we were in the middle of those mandated studies. Within a few years the Department of Interior would tragically recommend to Congress that the entire 1002 area be opened to oil and gas leasing.

At Beaufort Lagoon it's warm and buggy, with temperatures in the 60s—a heat wave for this arctic coast. The wildlife spectacle continues. Thousands of caribou mass together near the coastline, grunting and bleating, seeking relief from the thick swarms of mosquitoes. On this unusually warm and breezeless day the mosquitoes are at their worst. Many caribou pour off the ragged edge of tundra known to the Inupiat as Navagapak, which means "big point." They jump five or six feet down to a narrow stretch of sand wedged between the lagoon and the tundra. Some caribou walk out on the ice to escape the insects. A few caribou lie like sunbathers on floating ice cakes. If only the wind would pick up to help ground the pests.

"The bou are bunching up," says Fran Mauer, a longtime biologist on the Arctic Refuge staff. This phrase I would hear frequently during the next couple of days. Based on past studies, biologists believe that caribou form the largest, and most dense, postcalving aggregations when the mosquitoes harass them during warm, calm weather. The herd begins to bunch up in groups of tens of thousands of animals. Census photos have revealed as many as 80,000 caribou tightly massed in one group. Occasionally the entire herd masses together before they leave the coastal plain on their southern migration. In 1980 the entire Porcupine herd, then numbering about 110,000, left the coastal plain in one enormous procession, covering a distance of about ten to fifteen miles in breadth and one to two miles long.

From a biologist's perspective this is the ideal time to census the herd. It is Dennis's job to locate the groups by tracking radio-collared animals with the antenna-and-receiver system carried on board his aircraft. Once he finds large "bunches," he radios the Beaver, an ADF&G single-engine plane equipped with a large army-surplus aerial-mapping camera mounted on its belly. Its nine-inch-by-nine-inch negative format makes it ideal for taking pictures of tightly grouped caribou. If all goes well, we should be able to census the herd over a period of several hours.

On this day the whole herd is clustered in about two dozen groups scattered along the Niguanak River on the coastal plain and in the Egaksrak Valley in the Brooks Range about fifty miles to the southeast. Biologists must locate and count the animals in an efficient, precise manner. Otherwise, they risk counting groups of animals twice. They also search the surrounding area for any groups that might not have radio collars.

"Let's go for it," says Ken Whitten, a caribou biologist with the ADF&G. Federal and state biologists frequently coordinate their efforts on biological research projects within the Arctic Refuge because the state division has many biologists who are experts in their respective fields and who have conducted many years of animal research. Until the passage of ANILCA, there were only a couple of biologists on the Arctic Refuge staff, and little biological work had been completed within the refuge. After ANILCA's enactment, several new biologists were added to the Arctic Refuge staff, and USFWS entered into cooperative study agreements with the state of Alaska so that the five-year coastal plain assessment could be completed utilizing the best resources available.

We wait at the airstrip while a segment of the caribou herd crosses the runway. Once they are well out of the way, Dennis takes off in the Cub with another biologist, while three other researchers and I pile into the Beaver. The Beaver, with its large condorlike wings, slowly lifts off the strip, with ADF&G biologist

Caribou on Beaufort Lagoon

Pat Valkenburg at the controls. The 1950s vintage plane lumbers along the coastal plain, giving the sensation that we are in slow motion.

Within several minutes Dennis radios Pat that he has located a large bunch to the west. Ken and I are in the back of the plane, and we prepare to lower the camera, which is about the size of a small copy machine, through a hole in the belly of the plane. Soon we catch up with Dennis and spot a huge group of caribou milling around a large river bar, like a swarm of bees on their honeycomb.

"I'd guess about twenty thousand or so," Pat says as he circles the herd at an altitude high enough, about fifteen hundred feet, so that the animals are not alarmed.

"I'm going to open the hatch, so stand back and hang on,"Ken warns as he begins to raise a two-foot-square section of the floor. As he carefully pulls open the hatch, I look through the gaping hole down to the tundra. It is large enough for me to jump through, a fact that momentarily brings my stomach into my throat.

Ken gingerly unpacks and checks the camera, then lowers it into the hatch, taking care not to lose his balance. The camera fits snugly in the hole, and its 50mm lens is capable of capturing thousands of animals in one photo. Once we reach the edge of the group, Ken starts pressing the camera button, following Pat's signals.

When we have completed the first series of photos, Ken pulls the camera out of the hatch, and I grab the paper towels. The Beaver is notorious for spitting oil onto the camera lens, so it's my job to wipe the glass clean after each series of photos. With the hatch open, Ken tilts the heavy camera back just enough so I can wipe off the oil. As I clean the lens, the tundra races below me, and I have visions of a big bump catapulting me through the opening toward some unsuspecting bull caribou. Headlines read: "Woman Falls from Beaver, Gored by Caribou."

Within several hours most of the aggregating groups have been located. Tens upon tens of thousands of animals are crowded so tightly together that we can't see the tundra beneath them. In the distance the groups look like huge dark shadows of clouds. All of the animals seek relief from the mosquitoes. Some congregate along the coast, others swarm over river bars, and some climb ridges in the foothills seeking cooler elevations and a breeze.

When we return to Beaufort Lagoon, thousands of caribou pepper the ice floes beneath our wing. Sometime this month the fragmented lagoon ice will melt completely, and the sea ice will move north from the sliver of islands, given the right wind and weather conditions. There are a few open leads forming paths through the sea ice. Most of the white, gleaming layer resembles a sheet of broken shatterproof glass. The pieces still hang together.

In a normal year the ice will move several miles offshore by August or September, allowing barges to reach Barter Island and deliver annual supplies for Kaktovik's two hundred residents. Kaktovik is the only settlement that lies within the boundaries of the Arctic Refuge, an area almost as large as the country of Austria.

Ice-free waters also provide passage for beluga whales, and for the endangered bowhead whales during their westward fall migration along the coast. Kaktovik residents take advantage of the open channels as boating routes for hunting, whaling, and fishing trips.

It is late in the day now, and I decide to take a tundra stroll along the coast, walking along that final fringe of the continent. The sun rolls along the northern horizon, partially obscured by a thick mass of clouds that billows above the ice. I guess, by the tundra's rich color, cooling temperatures, and northern sun, that it is close to midnight. There is no need for clocks here, where days never end. You can read a good book at midnight if you want, but you probably won't unless it's storming and you're stuck in a tent. There is too much to see.

In the midnight sun arctic colors are at their richest; the low-angling light paints the tundra shades of amber and gold, and the sky rose-petal pink, saffron, and salmon. The light is soft, stunningly clear, as if the sea ice has pulled each ray from the sky, filtered from it any impurities, and thrown it back to the sun. Colors on the tundra are vivid, sharp. Dwarfed wildflowers, snug within the sedges and grasses, seem to jump out at you in their lavenders, pinks, and whites.

The land on top of the world magically glows at midnight, like flushed faces next to a winter's fire. Yet the level of radiance is more diffused in the Arctic, reaching the innermost petal of the tiniest wildflower, casting two-foot shadows beyond emerald sedges, bringing out the snow-whiteness of an old caribou antler.

Beyond the edge of moss and lichens, an eroding bank of ground and protruding roots falls three to four feet to the narrow shoreline, then to the lagoon with its giant puzzle pieces of floating ice. I walk down the tundra bank onto the narrow stretch of beach, which is covered with caribou tracks. With cooler temperatures the caribou have moved inland, leaving white tufts of their bleached, winter's fur scattered amidst the tracks. I pass several places where the caribou have bedded down in the moist sand, leaving shallow depressions, some with shorebird tracks scurrying through them.

A few hundred yards offshore I spot a Pacific loon, formerly called the arctic loon, diving for fish in a large open stretch of water amidst the floating ice. I have always loved the way loons proudly hold their heads, with their beaks tilted upward slightly, as if they are gazing at clouds. Then there is the loon's ancient, mystical call that seems to echo within itself before resounding across the water. It is a call that makes you stop in your tracks and ponder. The haunting voice of the loon seems to ask some unknown, ageless question.

Four species of loons (Pacific, common, red-throated, and yellow-billed loon) utilize the coastal lagoons and tundra ponds within the Arctic Refuge. The coastal lagoon system offers rich habitat for the fish-eating loon. The lagoons are protected by the barrier islands and are regularly fed with fresh nutrients from the many rivers and creeks that flow from the Brooks Range. Such waters offer ideal habitat for the prolific arctic char and many other species of fish, including the arctic cisco, flounder, and cod.

The Pacific loon winters along the Pacific coast and in Mexico's lagoons and estuaries. Several years ago Dennis and I wintered in San Ignacio Lagoon, on the west coast of Baja California. We were there to study the gray whales in one of their major breeding and calving areas. During our stay we spotted many of Alaska's migratory birds, including loons, at their winter home in the milder Pacific waters. I marveled at these northern migrants who had flown so far, and wondered if I had seen any of them in the Arctic Refuge only a few months before. Those south-of-the-border winged friends were kindred spirits.

Continuing along the beach, I notice a set of fresh arctic fox tracks weaving through the caribou hoof prints. I follow them along the shore, then up a small ravine onto the tundra. I walk over to a small pingo, a mound with an ice core that is forced up by frost action. It has been used as a perch by a snowy owl, and scattered on it are a few digested hairy pellets and one small white wing feather. Owls, as well as other birds of prey, regurgitate the hair and bones of their prey rather than passing them through their digestive tracts. I unravel one of the pellets and find a vole's tooth wrapped in its own fur.

Snowy owls are one of the few birds who live on the coastal plain year-round, although some of them will migrate into more southerly areas within the refuge. Because of their year-round white plumage, and the fact that they are one of the larger owls, they appear enormously striking against the tundra. They can be spotted at great distances during summer, like bright white sails at sea.

I decide that the shoulder of the pingo is a good spot to glass for the fox, owl, and other birds. Sitting down, I'm surrounded by several round cushions of moss campion. Each round cushion of mosslike leaves has the color and texture of a putting green, as though someone has manicured it. Springing from these mounds, one to two feet in diameter, are scores of delicate pink flowers rising a breath above the bright green cushions.

It is never silent on the arctic tundra during summer. Perhaps still, but never completely silent. Yet you always sense there is a ubiquitous curtain of silence shadowing all of the Arctic's perpetual voices: the lilting song of the Lapland longspur, the peeping sandpipers, the jaeger's cry, the loon's mystical call, the grunting of thousands of caribou. The arctic music is as constant as the twenty-four-hour daylight.

Across the lagoon I spot a large raft of old-squaw ducks,[1] their slender tail plumes backlit by the polar sun. Of the thirty-two species of birds that have been sighted within the Arctic Refuge's coastal lagoons,[2] the gregarious old-squaw duck is the most frequent visitor. In late July and during August approximately thirty thousand of these white-faced, black-breasted ducks congregate in the lagoons while they molt their wing feathers and begin staging for their fall migration.

Scanning the tundra with binoculars, I see the elegant long necks of a pair of tundra swans, formerly called whistling swans. Their graceful white necks and heads are held high as they paddle slow timeless circles in a tundra pond. Their coal black beaks, level with the Beaufort Sea, look as though they are balancing something on them. By now they must have hatched their young, and I wonder if the chicks are still in the nest.

The tundra swan is the largest breeding bird that utilizes the coastal zone of the Arctic Refuge. They mate for life and build large nests that are often reused. They usually nest near tundra lakes or ponds, often near river deltas. In recent years an average of 360 adult swans have bred within the refuge. After two or three chicks are fledged, and the swans complete their wing molt, they fly east to Canada, joining other segments of the population along the McKenzie River delta.

From the delta, the tundra swans follow the McKenzie River south across Canada's interior, and they eventually arrive at their wintering grounds along the eastern seaboard. For many years the majority of the tundra swan population wintered in Chesapeake Bay, until the bay's pollution level and subsequent loss of feeding habitat drove them south to coastal waters along the Carolinas. The Atlantic population of tundra swans now numbers about eighty thousand birds.

I'm about to leave my comfortable seat on the pingo when I spot the arctic fox loping toward me along the edge of the tundra. I sit still so as not to alarm the fox, hoping it will continue past me so I can get a close look. The fox lopes to within ten yards of me, pausing on the opposite side of the shallow ravine, its paw resting on a tussock. It has lost its white winter coat, and sports a new healthy-looking coat that appears soft and golden, blowing slightly in the trace of breeze.

The fox stares at me for what seems like forever, although probably only a minute or two. Again there is the eye contact; with this wild fox, its eyes are less trusting, more territorial. I'm not as comfortable looking into the eyes of a fox as I was with the caribou calves. I get the distinct feeling that it doesn't want me here. Perhaps I'm near its den.

The fox starts barking and howling at me, more high pitched than a wolf, much like the tone of a coyote. In my mind I say I'm sorry, I mean no harm. I stand up and slowly start to walk away. The fox still howls at me. I take a few steps then turn around. "I'm leaving now. It's okay," I softly speak to the fox. It stops barking and looks at me puzzled. Then it slowly walks away, down into the ravine.

I walk back to join the others, thinking about all I had seen in one day on the arctic tundra. This is truly the most extraordinary of ancient birthplaces. There is a sacred quality about this ground, this tundra that so exquisitely sweeps to the Brooks Range, where so many diverse species make their annual pilgrimage, traveling thousands of miles, to breed, to give birth, and to nurture their young. It is America's greatest wildlife mecca.

There is something inherently special about places where humans and all members of the animal kingdom bear their young. Such places are looked to with reverence, renewal of spirit, and celebration of life. In the human world we cherish the miracle of the birth of our children. We treasure our newborns, celebrate birthdays, and remember our roots as we grow old.

My grandmother once took me to a patch of ground, a part of her old Oregon homestead, where she and her father and her grandfather had all been born. They had begun their lives in the same farmhouse, which had long ago burned to the ground. Grandmother was flooded with memories and tremendous emotion as she walked in circles looking at the barren ground on a hot June day. The surrounding homestead lands had changed little, unlike other old places, which might be paved over or turned into shopping malls. She did not have to face that sort of sadness, as many others might have.

Many people celebrate the birth of Christ and recite the nativity story to their children. Thousands of people make their annual pilgrimage to Bethlehem to worship and give special recognition to Christ's place of birth. We celebrate and pay tribute to our great presidents, usually by recognizing their birthdays and enjoying a day off work. Our spirits are always uplifted with the birth of spring, a time of rejoicing.

Within the animal kingdom, we marvel at the salmon who travel great distances upstream to deliver their eggs, complete their life cycle, and die; or at the gray whales who travel thousands of miles from the Arctic to breed and deliver their calves in Baja's lagoons. On the coastal plain hundreds of thousands of birds and caribou have conducted their exhaustive annual migrations for millions of years to breed and bear their young on this northernmost stretch of plain. We should treat this special place, and all animal birthplaces, with the same kind of reverence and recognition that we give to birthplaces of our own species. Should we consider allowing the oil industry to exploit this ancient birthplace within America's only Arctic Refuge?

I stumble upon an old rusted fifty-five-gallon oil drum, a 1950s discard from the DEW line site. It is partially smashed, crinkled, and embedded in the tundra. Although most of the old drums east of Beaufort Lagoon have been recovered and removed by USFWS, there are still thousands scattered along the coast to the west.

I stare at the old drum for some time, picturing how oil and gas development would irreversibly destroy the wilderness before me—hundreds of miles of roads and pipelines, scores of drilling pads and rigs, thousands of oil spills, as in the case of the Prudhoe Bay fields. There would be spillage of toxic drilling muds into the wetlands, piles of abandoned industrial junk, and other pollutants: garbage, traffic, and noise.

The Arctic Refuge contains the only sliver of arctic coastal plain habitat—roughly 150 miles of coast, stretching an average of 25 to 30 miles inland—that is protected within a conservation unit. The remainder of Alaska's arctic coast, some 1,000 coastal miles, has already been dedicated for oil and gas development.

In the coming years the oil industry will endlessly lobby Congress, with promises that it will develop this coastal plain in an environmentally sound manner: build pipelines and roads so that caribou can cross them, use less toxic drilling muds, clean up their many oil spills, curtail development activities during the caribou calving season. Promises, promises.

In light of the March 1989 Exxon Valdez disaster—the largest oil spill in U.S. history[3]—where thousands of birds, otters, and other animals suffered and perished because of gross negligence and inadequate contingency plans on the part of the oil industry, it is unconscionable to accept industry's promises, particularly when the nation's premier wildlife refuge is at stake.

3

Snow Geese, Polar Bears, and the Great White

BY THE MIDDLE OF AUGUST, the vast majority of the Porcupine caribou herd has left the coastal plain, streaming through the Brooks Range, and dispersing into their more southerly winter range. The lanky calves have grown strong and stocky from their mothers rich milk coupled with two months of summer foraging. By November an average calf will have increased its body weight six times, and will usually have been weaned. Adults have also built up their fat reserves in preparation for the breeding season and long winter.

Nearing summer's end, the shorefast ice has moved away from the coast, creating an open passage for the bowhead whales who are about to begin their westward migration from the McKenzie River delta. Kaktovik whaling crews prepare their boats and equipment for the annual fall bowhead hunt. Whale hunting usually occurs within ten miles of shore, but sometimes as many as twenty miles offshore.

Alaska's subsistence whale harvest is jointly governed by the International Whaling Commission and the Alaska Eskimo Whaling Commission. The international governing body sets the annual harvest limit, and the Eskimo Whaling Commission determines how to divide up the quota between Alaska's communities. In recent years Kaktovik has been allowed two or three whales.

Whale meat and *maktak* (the skin and blubber) is the mainstay for Inupiat people. It is estimated that an average whale provides some twenty thousand pounds of meat and other edible portions, which is shared regionally among northern coastal communities such as Kaktovik, Barrow, and Wainwright, and inland villages, such as Anaktuvuk. Each year some villages are successful in obtaining a whale, while other communities are not. Thus, the regional sharing of the bowhead and other game, such as caribou, has been a common tradition.

While whaling crews are in search of the bowhead, other Kaktovik residents boat along the coast to hunt caribou, often members of the Central Arctic herd. Although the range of this 18,000–member herd is generally centered near the

Prudhoe Bay oil fields, the herd also utilizes the western end of the refuge's coastal plain, overlapping with the Porcupine caribou herd's range.

Many breeding birds have departed on their southbound migrations. Some, like the Lapland longspurs and snow buntings, follow river valleys through the Brooks Range, migrating southeast through interior Alaska and Canada. Others, like the tundra swans, wing their way east to the McKenzie River, joining Canadian breeding populations, then move south. Some shorebirds, such as the phalarope, follow the ragged edge of the arctic coastline west, then south along the Chukchi and Bering Sea shorelines, and on down the Pacific coast. A few species, like the arctic warbler and northern wheatear, continue west to Siberia, into Asia, and in the case of the wheatear, as far as Africa.

As inland bird populations decrease, there is a tremendous convergence of shorebirds and waterfowl in the lagoons and tundra ponds along the refuge coast as they prepare for their migrations. Birds who may have nested in upland tundra zones, like the lesser golden plover[4] and pectoral sandpiper, flock together near the mouths of rivers. Old-squaw ducks continue to gather by the thousands in the lagoons. Black brant congregate near river deltas, feeding in low-lying marshy areas. Flocks of dunlins and long-billed dowitchers gather on the mudflats and in tundra ponds. Some of these flocking shorebirds and waterfowl have nested within the Arctic Refuge, while others have traveled west from their breeding grounds in Canada.

Just why so many shorebirds and seabirds converge along the arctic coast toward the close of summer is somewhat of a mystery. One theory suggests that the birds group together, like the caribou, in preparation for their southern migration. Another theory suggests that the quality of the food source along the coast may be a key factor.

Philip Martin, a Fairbanks ornithologist, studied bird use of coastal habitats for two summers at the Canning River delta within the Arctic Refuge. He theorizes that many probing shorebirds may flock at river deltas to feast on protein-rich insect larvae in the numerous tundra ponds and on marine invertebrates in the mudflats. Many high-latitude insects, like the midge, have multiyear life cycles because of the extreme low temperatures and brief summer growth period. They simply don't have enough time to develop their wings and take to air within a one-year period, as is the case with insects in more temperate zones.

This means that a greater supply of insect larvae is available to birds in the Arctic during the latter part of summer. The farther north a bird flies, the more insects it will find in the multiyear larval stage. A shorebird can nest and feed on larvae and hatched insects in the inland areas of the refuge, then move north to the coast, where insect development is delayed because of the colder summer temperatures of the marine environment. There is a guarantee that more larvae await them.

By September the coastal plain has received its first dusting of snow, with temperatures regularly dropping below freezing. Night has returned. Tundra ponds grow still and are sheathed with surface ice. The mosquitoes have vanished. Blueberries have frozen and turned mushy. Fish, such as the arctic char, have migrated up several of the coastal plain rivers to spawn and seek deep overwintering holes.

The tundra is briefly drenched with shades of crimson, orange, and yellow during the fleeting fall. Rivers of golden willows meander across the sweeping plain, fingering their way into the Brooks Range. Autumn does not linger. It lasts about as long as a five-minute intermission during a two-hour movie.

The sun shines on the coastal plain for only twelve hours a day, compared to twenty-four hours in June. The spectacular northern lights paint the night sky with curtains of shimmering light curving around the pole like a halo. The magical lights, the first snowstorms, the migrating bowhead whales, return with another new visitor—the snow goose.

Beginning in late August and through the month of September, lesser snow geese arrive on the coastal plain by the thousands. They fly from their summer nesting grounds on Banks Island, about 450 miles to the northeast. As many as 325,000 lesser snow geese fly to the coastal plain each year to prepare for their long southern migration.

The coastal plain of the Arctic Refuge and adjacent Northern Yukon Park in Canada provides a critical pit stop for the birds. Here the snow geese bulk up their fat reserves for their long southbound flight, increasing their body weight by 20 percent. USFWS has calculated that adult snow geese spend 60 percent of their time feeding while they are on the coastal plain; juvenile geese spend close to 80 percent of their time eating.

Like the caribou, snow geese are attracted to the high food quality found in a different species of marshy cotton grass, *Eriophorum angustifolium*. The cotton grass makes up more than 80 percent of the geese's diet during their stay on the coastal plain. The most nutritious part of the cotton grass lies in the plant's stem base and bulb, which the snow geese pull out of the ground. Snow geese are ravenous when they arrive at the refuge, largely because during the nesting and brooding period the adult geese eat very little. Breeding grounds are often crowded, and there is great competition for available food. It is believed that most of the snow geese would be unable to complete their migration to California's Central Valley and other points in the western United States without the coastal plain stop.

What's more impressive is the fact that the snow geese fly several hundred miles out of their way to feed on the coastal plain of the Arctic Refuge. They feed over a period of several days, then retrace their route back to the McKenzie River to join the rest of the Banks Island population, a total of approximately four hundred thousand to six hundred thousand birds. One theory to explain why a large proportion of the snow geese detour to the Arctic Refuge coastal plain is that the competition may be greater for ideal feeding sites farther east.

While the geese bulk up on the coastal plain, they are extremely sensitive to aircraft disturbance and other noises. If the Arctic Refuge coastal plain is opened to full oil and gas leasing, as proposed, development-related disturbances could displace snow geese from as much as 45 percent of their preferred feeding habitat, as noted in the Department of Interior's 1002 report to Congress. It is theorized that such disruption and loss of habitat could seriously affect the geese's desired weight gain, potentially causing higher mortality during their strenuous southern migration.

The snow geese are the last of the coastal plain migrants to make their exit. Their departure marks the end of the field season for biologists, and a time when only a few of the most well-adapted beings remain on the coastal plain. Of the large mammals, only hibernating female polar bears and the muskoxen stay on the coastal plain through the long arctic freeze.

Sometime during the month of October, after there is sufficient snowfall, pregnant female polar bears seek out den sites on the coastal plain. They have spent the last six months living out on the pack ice, surviving on ringed and bearded seals. During a single year, it is estimated that many polar bears travel throughout fifty thousand or more square miles of sea ice, an area nearly as big as the state of Washington. The majority of the females den out on the ice, some traveling great distances toward the North Pole. One female polar bear was located in a den five hundred miles north of Barrow.

It is estimated that 25 percent of denning females within the Beaufort Sea polar bear population, numbering about eighteen hundred animals, choose to den on land or landfast ice, rather than on drifting pack ice.[5] Perhaps because the refuge offers better habitat than elsewhere on the North Slope, the coastal plain of the Arctic Refuge has the highest concentration of denning polar bears on the entire North Slope of Alaska. In recent years USFWS researchers have located most of Alaska's land-based denning sites within the Arctic Refuge. The topography of the coastal plain is generally more rolling than North Slope regions to the west. The leeward sides of bluffs and hills, with adequate snow accumulation, provide good denning sites.

Arctic fox

Another reason polar bears may frequent the Arctic Refuge coastal plain is that the zone between the active drifting pack ice and the landfast ice, where seals are more concentrated, is closer to the Arctic Refuge shoreline than to points farther west. The pack ice zone can be found within ten miles of the Arctic Refuge coastline, whereas it is approximately thirty miles north of Barrow. Polar bears have easier access to denning sites on land within the refuge.

Polar bears breed in the spring, but the fertilized egg drifts in the fallopian tube until the female begins her denning cycle. In early November the fertilized egg implants in the uterus, and a very short gestation period follows. Polar bears deliver tiny, helpless cubs—commonly two—around the first of the year; they weigh an average of 1 to 1.5 pounds each. The cubs grow rapidly in their snow caves and weigh about 25 pounds by spring, when they are ready to emerge from the den with their mothers.

Maternal polar bears are especially sensitive in the winter. They are quick to abandon their dens and abort their cubs if they are disturbed. During the winter of 1982, several oil exploration seismic crews were out on the coastal plain making checker board tracks with heavy equipment and shooting dynamite charges into the tundra to assess the coastal plain's oil and gas potential. When one seismic train ventured too close to a polar bear den, the female abandoned the site. She presumably either aborted or abandoned her cubs. Her den was one of fifteen that have been located on the coastal plain in recent years. The total number of polar bear dens on the refuge is unknown.

There are about forty thousand polar bears worldwide that wander below the North Pole. They are not considered "threatened" or "endangered," but they have been classified as "vulnerable" to human disturbance. An international polar bear treaty has directed five nations to conduct polar bear research programs, regulate hunts, protect females and young, and protect polar bear ecosystems. As a result of the 1976 treaty, some countries, like the Soviet Union, have banned all polar bear hunting. Other countries, such as the United States and Greenland, restrict the harvest to subsistence hunters.

Through the heart of winter, polar bear mothers spend all their time either nursing their cubs or sleeping within their snow caves. While much of the coastal plain may look wind-swept, barren, and somewhat lifeless during the winter months, it continues to offer a birthplace for polar bears and smaller mammals, like voles and lemmings, beneath its white crust of snow.

If the mantle of snow were transparent and you could look through a window into a cross section of tundra, you'd see a vibrant world. Through the shallow roots of the mosses, lichens, and other dwarfed plants you would see a maze of small tunnels zigzagging between the tussocks only a few inches beneath the ground. Beneath your feet you might see brown and collared lemmings, or tundra voles, scurrying through the corridors to feeding sites. They might be nibbling on the stem bases of the cotton grass, making use of the nutritious plants the way snow geese and caribou do during summer months.

Beneath "the great white," a phrase Dennis often uses when flying over the coastal plain in winter, these rodents can multiply by the thousands. Their reproductive success beneath the snow has a great bearing on the survival and reproduction rate of the short-tailed and least weasels, foxes, snowy owls, jaegers, and other birds who prey on rodents.

Through winter's window in the tundra you might also find arctic ground squirrels hibernating within pingoes or along river bluffs. The arctic ground squirrel is one of few arctic mammals that have evolved to sleep through the cold of winter. Sleeping is not an easy task. To survive the winter, the squirrel must regulate its body temperature so it will not freeze. This means the squirrel must locate its burrow within a very narrow band of soil beneath the tundra's surface. If the squirrel's den is a little too deep or too shallow, it will probably die from burning up its fat reserves if too warm or from freezing its body parts if too cold. The squirrel has the evolutionary advantage of not having to obtain food in winter, but hibernation has its risks.

In the middle of winter, I sit snug within a small cabin on the south slope of the Brooks Range and think of the arctic ground squirrels snoozing beneath the snow on the coastal plain, and polar bears and lemmings delivering their young of the year beneath the great white. Then there are those few hardy, well-adapted species, like the muskox, who survive the winter on the wind-lashed surface of the coastal plain.

I look through the log cabin's ice-encrusted window into winter's dark stillness, listening to the candle flame flicker from a cold draft slipping in between two logs that are in need of chinking. Lake ice beyond the cabin expands and cracks in the thirty-below temperatures, sending high-pitched wails into the night, like songs of the humpback whales. Within our niche we need another log on the fire. I bundle up and decide to walk outside to listen to the lake's voice and watch the aurora borealis paint the sky above the Brooks Range.

Outside it is so cold that I can feel the moisture in my nose crystallize as I breathe the frigid air. The silence is as deep and overpowering as the cold. Other than the sound of my breathing and the occasional eerie wailing of the lake ice, there is perpetual stillness. Above me the northern lights dance across the sky in undulating waves of blues and greens, as though someone has a paintbrush and is stroking the sky in free sweeping movements. Every so often the paintbrush brings a long stroke of dazzling color toward the earth, and for a moment I think the ribbons of light might touch the lake. The colors drip in suspension for seconds and then vanish to the heavens as though the painter has inhaled them.

Looking north beyond the lights and shadowed mountains, I think of all the arctic animals that are surviving out there in the deep freeze. Beneath the shimmering aurora are thousands of caribou cratering through the snow for their lichens, wolf packs following the caribou trails, and ptarmigan burrowed in a drift of snow, waiting for the dawn. Somewhere beyond the mountains is a polar bear in her snow cave giving birth to her cubs in the so-called dead of winter.

It is not so dead. The earth's plant life is in a dormant state, and many of the limited number of species who live in the Arctic have migrated to more southerly latitudes. But there are those few well-adapted species that continue to survive, year after year, through our "dead" of winter. They have successfully done so for centuries.

My nose borders on frostbite, so I grab a log for the fire and retreat to our warm niche. To survive in these latitudes I must wear several layers of clothing and regularly fuel the fire to maintain my 98.6-degree body temperature. I have not acquired a specialized coat of fur, like the caribou, muskox, or polar bear. I can't hibernate or thermoregulate the flow of my blood to my extremities the way other arctic mammals do. As a bare-bodied human, I would die in the Arctic in a few minutes without the necessary amenities.

Inside the cabin, warm once again, I realize that this Arctic Refuge out our back door is a special place in winter as well as in summer. I marvel at those few species who have successfully evolved to live in this northernmost landscape. Some not only survive through the long winter, they also reproduce beneath the snow, ice, and whip of polar gales. There is deep satisfaction in knowing that such wild habitat exists in its natural, unadulterated state to support these high-latitude evolutionary achievers. It is a sanctuary that belongs to them. I am only a fortunate visitor.

4

Marsh Creek to the Sadlerochit Mountains

IT FEELS LIKE WINTER ON JUNE 8, in the back of Walt Audi's Super Cub. Even with long underwear, woolen pants, a turtleneck, down vest, and windbreaker, I'm still freezing in the back seat. The patchy fog along the arctic coast and a biting polar wind funnel up my leg through a crack between the door and the fuselage. Arctic fog and wind cut to the bone, especially if you are lean. Today I wish I were ten pounds heavier.

The Cub putters just above the swirling fog that fingers inland from the arctic coast for several miles. There is not a hint of green from the air, and the winter-brown coastal plain is blotched with snow. It's hard to imagine all the hundreds of thousands of birds below us, out on the tundra, seeking nest sites beneath snow-free tussocks. From the air the landscape looks too foreboding, too cold a place for a bird to want to lay her eggs. Yet birds have successfully dropped and hatched their eggs here for millennia, just as the caribou have successfully dropped their calves.

I'm a day late because of weather delays. Ave Thayer, the former manager of the Arctic Refuge, is waiting for me at another abandoned DEW station located on Simpson Cove, about thirty-five miles west of Kaktovik. He is conducting a wilderness assessment of the coastal plain, another review mandated by ANILCA. I've volunteered to assist him on a two-week trip that will take us from Simpson Cove to the foothills of the Sadlerochit Mountains, about twenty miles inland, following the Marsh Creek and Katakturuk River drainages. I feel fortunate to accompany such a knowledgeable individual, a person who grew to love the refuge while managing its lands for twelve years.

Flying along the ice-choked shores of Camden Bay, just west of the Sadlerochit River, I spot some old sod-and-driftwood house ruins that were occupied by several Inupiat families in the 1920s. Inupiat residents refer to this coastal zone as Aanalaaq, which means "at the head of the bay." The Ologak family whose descendants live in Kaktovik, once lived and herded reindeer at Aanalaaq because the surrounding low-lying hills offered good visibility for the herders, and desirable habitat for the reindeer.

Reindeer herding was first introduced in Alaska during the 1890s, when several herds were brought to Alaska from Siberia. The government-sponsored plan was conceived by Alaska's superintendent of education, Reverend Sheldon Jackson. Jackson believed that reindeer herding would provide a stable source of food for Eskimo communities, considered impoverished at the time. During the 1920s and 1930s, the Ologaks were one of several families who herded reindeer on the coastal plain.

Each spring, during calving season, the Ologaks would take their reindeer into the foothills of the Sadlerochit Mountains, looping back to Aanalaaq later in the summer. It was difficult to keep track of the reindeer, and wolves often attacked them. When the mosquitoes were at their worst, grease from seal fat was reportedly applied to both the herders and the reindeer to help keep off the pests.

Reindeer herding gradually died out on the coastal plain in the mid-1930s after Eskimo residents and the reindeer experienced two extremely severe winters. During the winter of 1935-36 many families were destitute because no game was available, and some died of starvation or tuberculosis. The U.S. Bureau of Indian Affairs recorded the following comments of one family living at the mouth of the Canning River: "Destitute. Straight flour, no food, for four days. Boys healthy and girls dying from slow starvation and tuberculosis."

Other families reportedly ate raw flour, the skins of kayaks, and seal boot leather to survive. Most of the reindeer, numbering as many as twenty-seven hundred in 1930, either starved to death because of deep and ice-crusted snow conditions or were killed by people for food and clothing.

In response to the food shortage, the government initiated a second herd drive in 1938. Herders drove three thousand additional reindeer from Barrow to the Barter Island area. Soon after the reindeer reached their destination, they did an about-face and headed back to their home range around Barrow. Unfortunately, most of the remaining local reindeer followed the Barrow-bound herd, marking the end of the reindeer-herding era within the later-designated Arctic Refuge.

As we fly over Marsh Creek I spot a couple of canvas wall tents pitched on a sandbar that extends into the creek's mouth. This spot is locally known as Kunagrak, named after an Eskimo family that lived and trapped here in the 1920s. Today this location is a popular campsite for Kaktovik residents who hunt waterfowl and fish in the neighboring creek known as Iqalugliurak, which means "little creek with lots of fish." This same creek is labeled on United States Geological Survey

(USGS) maps as Carter Creek, named after a prospector. I much prefer Inupiat names since they usually tell a story about the natural features of the place.

Just beyond Marsh Creek is another one of those strange three-dimensional DEW complexes. On the treeless tundra the metal boxes stand out like semi-trailers flung on an open desert. The Simpson Cove and Beaufort Lagoon DEW stations are the only two abandoned sites along the refuge coast. The only active DEW station, manned with thirty to forty personnel, is located on Barter Island next to Kaktovik.

Walt slips down through a thin veil of fog to a dirt strip that lies between the buildings and the cove. I can see tall Ave, hands in his pockets, looking like a pole without its windsock, standing at the end of the runway. Walt smoothly touches down on the plane's oversized balloonlike tires that are designed to cushion rough landings. He wheels over to Ave, then shuts off the engine, which is unusual. On previous landings, Walt has usually left the propeller whirring to blow the mosquitoes away while he unloads the plane. The bugs aren't out yet, so off goes the engine.

A pair of snow buntings scurry under the wing of the plane, as we step onto the strip. They are reportedly one of the few perching birds that seek structures like the DEW line buildings to nest in and around. Arriving on the south slopes of the Brooks Range in early spring, usually late March, the snow buntings are considered to be the first spring migrant to return to the Arctic. They mark the beginning of the procession of many flocking migrants, such as the white-crowned sparrows, redpolls, Lapland longspurs, and shorebirds and seabirds that soon follow them. Most of the buntings nest in the Brooks Range in cliffy areas, but a few stray out as far as this northerly DEW line site.

Soft-spoken Ave, his windbreaker soaked from the drizzle, greets us as we unload my gear from the plane. Walt decides to take off quickly before he gets enveloped by the fog. He lifts the Cub into the icy breeze toward the North Pole, then banks to the east, slipping between the fog and the coast, puttering his way back to Kaktovik. Ave and I listen to the plane fade away beyond the fog-shrouded bay, then head for one of the DEW line buildings to get out of the drizzle.

We walk into a corrugated metal vault that is the coldest, darkest structure I've ever entered. The windowless noninsulated walls offer little protection from the arctic cold. The air temperature seems colder inside than out, but at least we are out of the drizzle and wind. Ave suggests that we sleep in the vault since we still have to sort out and package food for the trip, and it's growing late in the day.

In the dim light we package food on an old table near the door. The temperature is in the 20s. By the time we finish, the tips of my thumb and index finger are nearly frozen together from sealing all the plastic bags. Ave brought all the food, except for my stash of m&m's and hot chocolate. As I look over the piles of dinners, lunches, and breakfasts, I realize that during the next two weeks I will be eating the highest starch, lowest sugar diet of my whole life. Ave is big on noodles and potatoes, and sugar simply doesn't exist. Unsweetened oatmeal in the morning, no hot chocolate, peanut butter without jam, and absolutely no candy bars.

After we eat a spaghetti dinner, I decide to go for a tundra walk east of the metal box. The drizzle has turned to a fine mist, and the sun is still hidden behind heavy, dense cloud layers. I intended to stride out and cover some ground with my belly full of noodles but instead become interested in all the bird activity in the immediate surroundings.

Dozens of semipalmated sandpipers are pairing up, carrying on their courtship displays. The males soar and hover several feet above the tundra to advertise their territory, crying a short "creeep." I closely watch one pair only a few feet away. The male lands next to the female and begins parading around her with his wing extended. Circling her with dangling wing, he looks like a matador trying to attract a bull with his cloak. After a few moments the pair dart off to the shores of a nearby tundra pond.

Lapland longspurs, perhaps my favorite tundra bird, seem to be the most abundant and vocal. Their song can lift anyone's spirits on the dreariest day. During the worst weather on the coastal plain you can always hear a longspur singing somewhere. Edward Nelson, who extensively studied the longspur, wrote that "they are so filled with the ecstasy of life in spring that they must rise into the air to pour forth their joy in singing. . . ." And they do pour forth the most joyful and heart-warming of tundra bird songs.

Numerous longspurs surround me, the males being the most conspicuous with their colorful head markings. The striking velvet-black face and breast, with a golden streak behind the eye and bright chestnut on the back of the neck, make them easy to spot. The males all carry on the same flight ritual, which I watch over and over again. With rapid wingbeats they lift about twenty to thirty feet above the tundra, pause with their wings upturned in a V-shape, then gently glide down to earth, singing a liquid melody in harmony with the uninterrupted descent.

Lapland longspurs are the most common breeder on the coastal plain, and they are found throughout the circumpolar realm. Unlike other passerines who may live in forested or riparian environments, the Lapland longspurs don't need trees or shrubs to establish their breeding territory. Instead they hover above their territory during the courtship period, as if they are momentarily perched on a tree. Longspurs build ground nests, often lined with ptarmigan feathers, on the warmer, southern side of tussocks.

What is interesting about the Lapland longspurs is their versatility. During the summer months you find them mixed with shorebirds, and you sometimes suspect that one day you'll see them wading in a tundra pond probing for insect larvae, alongside a red-necked phalarope. Their sparrowlike beaks aren't designed for probing in the mud; however, they are well programmed to eat a wide variety of insects during their stay on the coastal plain. Unlike other nesters, who may prefer a certain type of habitat on the coastal plain, be it marshy, well drained, or riparian, the longspurs reproduce everywhere.

It is early June on the coastal plain. The cold, damp air seeps through my long underwear right to my bones. My chilled skin, with all the goose bumps, takes on tussock relief. I dream of sunshine, summer solstice, and my garden vegetables growing by inches in Fairbanks. I picture myself in a T-shirt, weeding the garden, sweating out a tennis match, drinking root beer floats.

Summer feels months away here. The temperatures have barely begun to climb above freezing. The ground has little spring to it since the surface of the frozen plain has only thawed a few inches. I walk between snow patches thinking of hot chocolate. Many of the tundra ponds are covered with ice, yet in the distance a tundra swan has found some open water. The elegant bird occasionally delivers a high pitched cry, perhaps calling for its mate, or maybe asking, "Where is summer?"

I walk back to the DEW station listening to the continual melodies of the Lapland longspurs mixed with the peeping sandpipers. I don't look forward to sleeping in the vault on a cement floor. Tomorrow we will move inland, away from man's ghostly structures, into the wilderness.

Morning found us both as cold as a couple of slabs of beef in a freezer. Neither of us had slept well in the vault. We are anxious to get under way, so we inhale some hot tea and oatmeal before it freezes in the cup, then shoulder our heavy packs. We try to figure out who has the heaviest pack and decide they are both in the fifty- to sixty-pound range.

Outside the vault a heavy low ceiling of clouds hangs like an immense wet sponge above the coastal plain and Beaufort Sea. Yet the clouds seem to have a blanketing effect. The temperature has climbed to forty-one degrees and the sea-ice breeze is light. I'm still damp cold to the bone, so I leave on my long underwear and wool pants. Ave does the same, and adds even more.

"How many layers do you have on?" I ask him.

"Seven," he answers, counting through a series of shirts and windbreakers.

Ave is a strong believer in wearing lots of layers that can be easily shed as he treks along. Or perhaps he thinks it makes his pack lighter.

We stride out wearing hip boots because of the wet, snowy terrain. Our feet and ankles would quickly turn to ice cubes in leather boots. We gain about two miles along Marsh Creek before the wind picks up from the north and a tongue of fog swirls toward us from the coast. Marsh Creek has only recently begun to break up. Several snow and ice bridges partially cross the narrow stream, and we see fragments of ice in the rush of water. Without a spark of sunshine the dark creek looks untouchably cold.

Marsh Creek is not particularly marshy as the creek cuts through gentle, undulating slopes. The creek's name is not derived from the landscape but, instead, from a prospector named S.J. Marsh. The lnupiat have always called the creek Nuvugaq which means "point that juts out in the water." The creek lies just east of a narrow point that arcs around Simpson Cove, known on maps as Collinson Point. Captain Richard Collinson was the British fellow who sailed along this coast in the mid-1850s in search of Sir John Franklin and his mysteriously lost expedition.

For a short time we follow the ugly tracks of a "weasel," a vehicle that was used briefly back in the early 1950s by DEW personnel. The obsolete, multitracked vehicle was about twice as large as a standard van, and ran on a V-8 engine. It was used in the immediate vicinity of the DEW station, and unfortunately left a series of narrow track scars across the tundra. Over the years the tracks have rutted five to six inches into the fragile tundra.

"You can see where thermokarsting has begun here," Ave says, pointing to fracture lines that run through the ruts.

Under normal conditions only the top two feet of the permafrost layer thaws during the arctic summer. This "active layer" of unfrozen ground is protected by the tundra's blanket of dense, dwarfed vegetation. If that shield of vegetation is removed or disrupted, abnormal thawing can occur within the permanently frozen ground layer. The permafrost band on the arctic slope reaches depths of approximately one thousand to two thousand feet, with an average temperature of about twenty-eight degrees.

Permafrost contains a high proportion of ice. If the ground is not insulated from solar radiation and other sources of heat, the ice melts and refreezes, and the soil settles, shifts, slumps, and collapses. Around Fairbanks you can see many poorly designed houses, built directly upon permafrost, that have buckled and sunk into the ground. The process is known as thermokarsting.

One of the most dramatic examples of thermokarsting is a section of road near the University of Alaska in Fairbanks. For many summers there was a twenty- to thirty-foot section of road, built on permafrost, that gradually would turn into a horrible dip by summer's end. The dip could easily knock off your exhaust pipe if you drove too fast. Researchers documented that the unstable ground sank at the rate of one-quarter inch per hour during August. For years highway crews would simply add more feet of asphalt to level out the dip, which only worsened the situation, adding more weight and heat to the already thawing permafrost.

Finally someone got wise and inserted thermal tubes on one side of the road, the same tubes that run along the trans-Alaska pipeline. These tubes help refrigerate the ground by allowing cold air to sink into the ground during winter months and by releasing heat to the surface during summer months. The engineering system, originally designed by an Alaskan inventor named Joe Balch, effectively helps stabilize the ground.

Ave suspects that the old weasel tracks, caused by summer driving on the fragile tundra, may continue to deepen over time. In the 1980s similar industrial tracked vehicles are generally prohibited from use during summer months on Alaska's North Slope. Yet there is still great concern over winter-only seismic traffic and the use of D-9 Cats on a landscape that receives little snowfall to protect its surface. Tundra wounds take many, many years to heal, if they ever do.

After a couple of hours of squishing across wet tundra, beyond the last weasel tracks, we decide to take a lunch break in the lee of a low creek bluff. Sitting on a tussock, I can't help but wonder whether we are resting on top of millions of barrels of oil. According to USGS reports, our lunch spot is located directly on top of the beginning of the Marsh Creek anticline, an upthrusted crust of earth with a supposed high potential for oil. I hope the reports are dead wrong.

In the coming months, during the winters of 1984 and 1985, oil and gas seismic crews would begin their checkerboard sweep of the coastal plain with their multiton Cat trains. They would evaluate the coastal plain, etching three-by-six-mile grid patterns across the thin band of snow cover, shooting explosives some eighty feet into the permafrost, and inevitably disrupting the tundra.

One of Ave's roles is to advise the USFWS on how winter seismic crews can assess the coastal plain with the least environmental damage, in an effort to preserve the plain's wilderness values. It is not an easy task given that the coastal plain receives little protective snow cover, and about 40 percent of the study area is made up of gentle foothills; it is a rolling plain, not a flat pancake. Such hilly topography, combined with the bumpy tussocks and multiple stream crossings, makes it almost impossible for the Cat trains to proceed without chewing up some of the tundra.

Wading across Marsh Fork

A rough-legged hawk soars above us while we eat crackers and cheese. Occasionally it delivers a high descending cry. The rough-legged hawk is one of nineteen species of raptors who live within the Arctic Refuge, and the one most commonly sighted. They, along with golden eagles, gyrfalcons, peregrine falcons,[6] and other birds of prey, nest in craggy areas in the mountains, foothills, and along river bluffs.

We are about fifteen miles from the Sadlerochit Mountains, which might be the home base for this hawk. By the signs of all the lemming and vole trails weaving through the tussocks, this is a good place to catch rodents. Perhaps we are sitting on one of the hawk's usual stops. Near our lunch site we find a vole that has been almost entirely consumed by some predator. Only the skin and part of the head remain.

After lunch we follow the creek for a short time and soon find ourselves surrounded by rolling hills sloping gently from the creek. The only hints of green along the creek are the unfolding feathery leaves of *Oxytropis*, which commonly grows along rivers and gravelbars. *Oxytropis* is one of the first flowering plants to emerge on the coastal plain. Gradually the creek valley narrows until we are forced to gain some elevation, cutting up a snow-drifted bank along the river. The river is too swift from spring snowmelt for us to wade through it.

Ave has the fortunate grace, or perhaps luck, to ascend snowbanks without sinking in too deeply. Twice I sink to my hips and, with the heavy pack weighing me down, have a heck of a time getting out. Somehow Ave manages to sink only to his knees, if at all.

Above the creek we find ourselves in terrain we label "tussock city." As far as we can see, there are slopes upon slopes, bumps upon bumps, of endless tussocks beneath the cold white dome of sky. Hundreds of thousands of the brown grassy heads pop up above the settled snow. It looks like the high-rise district. I've never seen so many miles of giant tussocks, in the two-foot range. These must be the type that some USFWS biologists, upset with administrative policies within the Department of Interior, have dubbed "Wattheads."

We decide to walk overland amidst the tussocks, trudging along snow paths, as wide as the soles of our boots, that weave around the mounds. Ave leads with a long stride gracing over the tussocks without stumbling, like a caribou. When you walk through tussock country, you must constantly concentrate on your footwork, or you're bound to trip and wedge yourself and your pack between the bumps. For many steps all I can see are the white snow maze, the grassy bumps, and Ave's black hip boots striding ahead with monotonous persistence. Snow, bumps, boots. Snow, bumps, boots.

Every so often we come across the delicate blooming purple mountain saxifrage and the emerging heads of woolly louseworts, two other early flowering plants. The dusty-gray furry heads of the lousewort remind me of the small clumps of pincushion cacti which grow in the Sonoran Desert. Momentarily I dream of baking in the sun in the Southwest, away from the snow, bumps, and boots.

The tundra is flooded with spring runoff. We can't walk more than twenty to thirty paces without crossing a rivulet of snowmelt. In a few spots we see mosses just beginning to green up by the water's edge. Because the plain has thawed only a few inches, the ground is firm beneath the melting snow, making the walking easier. We still have to wind tediously through the tussocks, but at least we don't sink into a swampy sponge, as will be the case when more thawing occurs.

After zigzagging our way through the tussocks, avoiding flooded areas, we spot a decent tent site that looks fairly dry and without the monstrous tussocks. We've probably walked about six miles as a crow flies, but I feel as if we've gone ten. We descend down one last snow-drifted slope into a broad ravine near Marsh Creek. I break through one of Ave's steps; my left leg plunges all the way in, over the top of my hip boot. I feel as though I've fallen into a crevasse.

With the pack anchoring me in, I have little energy to ease out my swallowed leg. My foot is apparently caught under a layer of crusted snow. After a great deal of pulling and twisting, out pops my leg and foot without the boot. I must look pretty helpless lying in the snow on my backpack like an arctic turtle, holding my foot up into the north wind. Ave backtracks and shakes the boot loose.

We set up the tent on a patch of relatively flat tundra not too far from the creek. The arctic blow is increasing, and we are anxious to put on dry socks and warm up in the tent. In no time we erect Ave's two-man yellow tent and begin to heat up water on his alcohol-burning stove. The simple things—dry socks, warmth, and crawling into my sleeping bag out of the wind—are a wonderful relief. This nylon shield of a home is like the Hilton to me.

I'm so tired after dinner that I decide to put off going out in that howling wind. I can't move out of my bag. Ave gets some kind of award for doing dishes in that frigid snowmelt water. As I drift off to sleep, I hear redpolls and Lapland longspurs singing between the gusts of wind.

Ave tells me in the morning that the tent flapped all night and the wind never let up. The noise didn't bother me; I slept soundly through all of it. We eat pancakes cooked on the silent-burning alcohol stove, which is fairly efficient, although it requires more fuel, in weight, than a gas stove. I'm accustomed to hearing the roar of my Mountain Safety Research burner, so I appreciate the silence, and the freedom to hear bird songs outside the tent, or a bear, if one strolls into camp. There is also satisfaction in defiantly burning alcohol instead of a petroleum product, given the oil industry's desire to rape the ground beneath our feet.

We break camp in what feels like a fifteen-mile-per-hour wind, gusting to twenty or twenty-five. The thermometer reads thirty-seven degrees. It's no fun swinging the pack on, with all the new aches and pains from yesterday. My hipbones feel as if they've been run over and grated by one of those seismic Cat trains. We head up Marsh Creek with the wind at our backs most of the time, wishing that our pack covers could convert to sails.

Along the creek we find more Lapland longspurs and spot a couple of yellow warblers in the willow thickets along the creek. Their bright golden breasts stand out against the budding willows like tiny crescents of sunshine. The water continues to run swiftly, and is slightly turbid. Because of the snow in the creek's valley, we decide to gain some elevation, returning to the slopes of tussock city.

As we tediously plod through the tussocks, following the ridgeline above the creek, I accidentally scare a mother willow ptarmigan off her nest, which is sheltered at the base of a tussock. Snug in the grassy nest are four rust-colored, black-specked eggs about two-thirds the size of chicken eggs. She flies off the nest for some distance, and I hurriedly leave, knowing the eggs need warmth on a day like today. Willow ptarmigan, and other nesting tundra birds are well camouflaged in their summer plumage. I'm rarely able to spot their nests unless I practically step on one.

The day passes slowly as we arduously trek through the tussocks, crossing snow patches and more runoff streams. The steady northeast wind erodes our faces and flattens the grasses and dwarfed plants to the ground. Again I expend a good portion of breakfast and lunch calories pulling myself out of those knee and thigh-deep sinkholes. I figure out why Ave doesn't sink in as much. The soles of his boots are much larger than mine. He can displace his weight better, the way a caribou does with its large hooves.

Most birds are apparently clinging to the tundra in the lee of the tussocks. The only birds who dare to wing with the wind are a few parasitic jaegers and a pair of rough-legged hawks. But we can still hear the faint cheery songs of the Lapland longspurs through the whipping rush of wind. It is as though they sing out of love for this polar region, regardless of the weather's adversity. Their song keeps me going and keeps my mind off my sore shoulders and lower back.

Later we give up trying to locate a dry, smooth spot for our tent. We have moved inland from Marsh Creek a few miles, and the relentless tussocks make it impossible to find a tolerable tent site. Instead we pick a solid, flat snow patch, knowing that we won't have to snake around the bumpy tussocks in our sleeping bags.

The temperature is thirty degrees, and the oppressive wind has not let up once for more than twenty-four hours. We guesstimate by our flapping tent and our ability to lean against the wind without falling that the arctic gale is steady at about twenty miles per hour, or more. Just before I fall asleep in the din of the flapping tent, Ave enters the tent looking windburned and half frozen.

"It's like Admiral Byrd going out to feed his howling dogs," he says with a chuckle.

Without a sense of humor we could really fall into a chasm of arctic depression out here in the miles of tussocks, beating wind and icy temperatures. Ave keeps it light. As we drift off to sleep he tells me that tussocks are really a sleeper and that they haven't hit the tourism market yet. He says that he'll capitalize on the tussocks, leading adventure trips across the coastal plain. Maybe with an overweight-clinic approach. He'll call his guiding service Tussock Tours.

"Just think," he laughs, "you could seat at least one thousand people on one acre of tussocks."

I drift off to sleep picturing one thousand people seated on one thousand tussocks, all watching thousands of migrating caribou stream by. An amusing image for my weary body.

The next day the wind continues to beat the tent and lash the tussocks. We decide to base on the snow patch for a few days and hope for some better weather before we break camp. There is also the chance that this might be the most level solid snow patch for miles.

With light daypacks we explore the undulating ridges of tussocks north of our camp. My calves are sore from yesterday's hiking, and I can barely keep up with Ave's long strides and quick gait. Before we have traveled very far, I spot a lone snowy owl lifting its strong, silent wings above a distant ridge. White bird upon white sky. Its ghostly image blends with the swirling low overcast that hangs above us. The bird vanishes quickly, like an illusion, into the white.

The few times I've seen snowy owls on the coastal plain I've been surprised by their large size and huge wingspan, which averages about five feet. Because their plumage is immaculately white, they appear striking against snow-free tundra, although today's white sky makes them almost invisible. The snowy owl is well adapted for winter survival, along with a few other conspicuous year-round residents, like the willow and rock ptarmigan and the raven. Snowy owls have the greatest plumage density of tundra birds and the lowest heat-loss rate, three to four times lower than that of the thick-feathered ptarmigan. Through the year snowy owls depend on rodents as their main source of food. In captivity they require four to seven lemmings per day, an indication of their dependence on rodents for survival. If rodent populations crash, snowy owls simply don't breed.

Later in the day we are forced back into the tent because of the foul weather. The wind is never ending, and the cloud ceiling has dropped lower, turning ominous. Looking out the door of the tent beyond our slope of tussocks, I see slate-colored clouds spit sheets of ice pellets across the gray-washed tundra. The storm is blowing toward us from the northeast, and within a few minutes our tent is besieged by sleet and hail. With the tent flapping and hailstones battering the nylon, we have to raise our voices to be heard through the din.

"You're just like a war correspondent writing under gunfire, with all that noise going on," Ave says as I work at my journal with frozen hand.

We start joking about the advantages of sitting in a tent at thirty degrees in a wretched arctic blow on the coastal plain. There is no chance a tree will crash upon us. The nearest spruce is about seventy-five miles away. No chance the tent will blow away since the stakes are frozen into the snow and tundra. No chance of mosquitoes or sunburn. All of these advantages can be incorporated into the Tussock Tours brochure.

The following morning we awake inside a snow cone. The tent is plastered with ice, crisp to the touch. The temperature is thirty-four degrees. I want to stay in the bag. Only nine more days until the longest day of the year, and we haven't seen the sun yet. The wind has tempered down a bit but still blows steadily out of the northeast. I wonder how the hundreds of thousands of nesting birds fared through the storm and how many lost their clutches of eggs. I would later find out that this particular month of June was, because of the weather, one of the worst on record for nest productivity.

Outside the tent tussock heads are snow coated for as far as the eye can see. Willow buds are sheathed in ice, and I wonder if they'll ever leaf out. A heavy white mantle of clouds and fog is suspended about twenty or thirty feet above the tundra. It appears to be slowly rising, as if the narrow cushion of air between tundra and clouds is palming the great table of white, thrusting it upward inch by inch, to the distant, hidden blue.

The wind has died down enough so we can hear the Lapland longspurs sing their soul-warming melody. Across winter's tundra another phantom: a snowy owl rises above the whitened plain into a cold fleece of sky. Its long arched wings beat slowly, persistently, as if the owl knows where it's going. As the owl grows more distant, the great white sky and earth blend together into one flat milky plane, absent any depth or horizon. The owl is the only clue that the plain ends somewhere and the sky begins. As it disappears into the great white, I wonder if it will catch a lemming or perhaps find a hint of blue sky.

With scud and drizzle much of the day, we spend most of our time in the tent reading, talking, writing, and occasionally cranking up the stove for some hot chocolate, or tea, in Ave's case. He reads *A Wilderness Bill of Rights*, by William O. Douglas, and I get more goose bumps while thumbing through *Minus 148°*, Art Davidson's account of the first winter ascent of Mount McKinley. Both books generate philosophical conversations about wilderness and wildlife and its incompatibility with industrial development.

Ave has seen considerable development and loss of wildlife habitat during his thirty-one years of USFWS fieldwork in Alaska. He was in the Prudhoe Bay area before and during oil development, and he has personally seen the land, lakes, and tundra ponds polluted with oil spills, drilling muds, camp sewage, and garbage. He has seen creekbeds destroyed for their gravel, the permafrost disrupted, and the numbers and diversity of wildlife reduced. Animals such as the wolf and grizzly bear have been poached, and eliminated in some areas, along the trans-Alaska pipeline and within the oil development area.

"I'm convinced that petroleum development here, on the coastal plain, is incompatible with the Arctic Refuge's purpose," he says.

The original 8.9-million-acre Arctic National Wildlife Range was established in 1960, under the Eisenhower administration, to preserve its unique wildlife, wilderness, and recreational values. Twenty years later, ANILCA added four new specific purposes for the establishment and management of the Arctic Refuge. The purposes are very clear in Ave's mind, and incompatible with oil development. In short, they are: to conserve fish and wildlife populations and their habitats in their natural diversity; to fulfill international treaty obligations, such as migratory waterfowl agreements and the Canada–U.S. Porcupine Caribou Herd Treaty; to provide an opportunity for local residents to continue their subsistence way of life; and to protect water quality and its quantity within the refuge.

"As refuge manager, I believed in following the letter of the law. In my view, I had no authority to interpret or compromise the laws which apply to the Arctic National Wildlife Refuge," he says in his soft but absolute voice.

Ave held a hard line when it came to questions over refuge development. In one case the Bureau of Land Management wanted to extend numerous oil lease tracts into the Arctic Refuge, slightly across the Canning River, which forms the western boundary of the refuge. The agency argued that if the tracts were squared off on the refuge side of the river, it would save on land survey costs, since they wouldn't have to follow a meandering river.

"There was quite a flap about those oil lease blocks. My answer was simple. Either the Bureau of Land Management would have to conduct a more costly meander survey along the river, or, better yet, square off the leasing blocks on their side of the river," he said.

Ave prevailed in that particular case, and in other instances. When the oil companies pleaded that they be allowed to remove gravel from the Canning River for construction of drilling pads, they were told no. When the industry wanted to set off a few seismic shots in the Canning River, Ave said absolutely not.

Ave's hard-line approach was noted at a USFWS administration meeting in Anchorage several years ago. A moderator asked several wildlife refuge managers how they felt the 1964 Wilderness Act should be interpreted. Managers were instructed to stand on the left side of the room if they felt they should ignore the act, in the middle of the room if they believed in compromising the law's provisions, or on the right side of the room if they believed in strict interpretation of the law. While most managers took their places in or near the middle of the room, Ave walked to the right-hand side of the room and stopped when his shoulder touched the wall.

In March 1988, Ave would testify before the U.S. House of Representatives Committee on Merchant Marine and Fisheries regarding the question of whether the coastal plain of the Arctic Refuge should be opened for oil and gas development. Again Ave would emphasize that there is no place for development within the boundaries of the Arctic Refuge. He would conclude his remarks by advocating wilderness classification of the 1002 coastal plain area, stating:

"We should leave the flowered meadows, the rounded hills, and clear brooks untracked, and leave the animals free to continue their ancient ways. We should not take nature apart. We should not take the Arctic National Wildlife Refuge apart."

Although Ave probably knows the refuge well, and has managed its lands longer than anyone else, his words sadly would fall on many deaf ears within the political arena. By the time Ave would reach Washington, D.C., many politicians would have already been wined and dined by oil industry representatives, and lobbied heavily by both industry and the Department of Interior, hearing such arguments as: the coastal plain is just a "flat, crummy place," and there is nothing unique about it; caribou and oil development are perfectly compatible; the industry has a clean track record at Prudhoe Bay; and the nation needs the estimated 200-days' supply of oil within the Arctic Refuge for reasons of national security, to balance the federal trade deficit, and, as some have ironically proposed, to export Alaska's oil reserves to foreign countries. Funny how all those industry arguments never mention the most important bottom-line word: profit.

Yet there would be those who would listen to Ave, those who would be sitting on the fence, those who had not bought industry's arguments, those who believed in the principles underlying the establishment of wildlife refuges within America. But would there be enough of them?

A few mornings later we are stunned to awaken in a tent not rattled by the wind. For the first day in seven it is dead calm, and the intermittent sleet, hail, and snow appear to have finally ended. I poke my head outside the tent, expecting dense fog and zero visibility. Instead, the fog is dissipating, and for the first time I can see the base of the Sadlerochit Mountains. Up until this moment the mountains were only a fantasy labeled on our topographic maps.

It's invigorating to see the beginning of some mountain relief after days of little visibility and endless tussocks. The entire mountain range gradually reveals itself as the curtain of fog lifts higher and higher, along with our spirits. The Sadlerochits are in the 4,000- to 5,000-foot range, but their snow-dusted faces appear massive, since we are only a few hundred feet above sea level.

"These Sadlerochits are as grand as the Himalayas," I tell Ave, while running for my camera.

The Inupiat name Sadlerochit means "area outside of the mountains." The Sadlerochit Mountains can be considered an extension of the Brooks Range, yet they are indeed the northernmost set of mountains, flanking the Brooks Range for about fifty miles, running west from the Sadlerochit River to the Canning River. From the air the Sadlerochits appear to be a separate range since they front the Brooks Range as an isolated set of mountains, stretching several miles farther north than any other portion of the Brooks Range. The Sadlerochits are young mountains and display the most recent signs of uplifting. One geologist pointed out to me that when you approach the Sadlerochits along the Marsh Creek drainage, you are really walking on top of the backs of new mountains that are rising beneath your feet.

There is something special about experiencing your first view of a mountain range, or an individual mountain, after it has been mysteriously buried in clouds for days. After great anticipation, there's a sense of magic when the mountains suddenly appear, as if they were just born. It is much like the magic of a child opening a favorite Christmas present after many days of waiting. The longer the child has waited, the greater and more meaningful the surprise.

I first saw Mount McKinley at close range in much the same way I saw the Sadlerochits. One overcast summer day I was hiking with a friend within Denali National Park, and Mount McKinley was entirely enveloped in clouds. After walking several hours through intermittent drizzle, and then a rain squall, a series of nature's phenomena occurred all at once. As the squall blew past us we found ourselves walking through a fine mist of shattered diamonds sparkling in a lone patch of sunshine. At almost that exact moment a moose suddenly appeared, as though it had fallen from the clouds. The "apparition" trotted by us, through the thin mist of glitter, across the shimmering green tundra. Within seconds the brightest rainbow appeared above the crest of a distant ridge. And beyond the rainbow, the highest mountain in North America began to reveal itself through the parting clouds. We saw the towering face of Mount McKinley for only a few minutes before it vanished into the clouds again. I will never forget those magical moments.

Today's first view of the Sadlerochit Mountains reminded me of the Mount McKinley experience. We had lived through a week of almost intolerable arctic weather, stumbling through the tussocks in the biting wind, sleeping in damp wool clothes and bags, cursing the sleet, snow, and flapping tent. The adverse conditions we had just experienced were all a part of the wilderness experience in the Arctic, and somehow it all seemed worth it.

It was worth it because of nature's rewards, both great and small: suddenly seeing this exquisite northernmost mountain range after days of poor weather conditions; humbly watching a ptarmigan tenaciously sit on her nest of eggs in the freeze of an arctic gale; hearing the song of the Lapland longspur penetrate through the relentless wind to warm the soul; joyfully living a day without the grating wind, sleet, snow, and so many extra layers of clothing; then celebrating the return of the perpetual arctic sun.

The marathon runners who push themselves to the limit of pain and endurance and finish the race physically exhausted achieve the goal of finishing the race and running their best. This is the runner's greatest reward. All of the pain, misery, and exhaustion is worth it.

The wilderness hiker is somewhat like the marathon runner. The hiker treks through miles of rough terrain carrying a hefty house on his or her shoulders, faces unknown weather conditions, and suffers at times from mental and physical exhaustion. But the rewards are there. For the runner it is the finish line. For the wilderness hiker it is the gifts of nature and a greater appreciation of life.

Without spending some endurance time on the coastal plain I would never fully understand or appreciate how enormously difficult it must be for a bird to successfully reproduce in the Arctic, particularly after having migrated thousands of miles. Nor would I fully appreciate the many small comforts that I take for granted in my daily life: a dry pair of socks, a warm bath, fresh fruits and vegetables, a candy bar. To struggle in wilderness is to better know myself and to better appreciate what is left of our remarkable natural world.

Several days later the wind and freezing temperatures have returned as we trek to the coast along the Katakturuk River. The Inupiat word means "river where you can see a long way." As evidenced by our route along the high bluffs above the broad, gravel-braided river, it is clear why the Inupiat chose the name. From the bluffs we can see for miles down the multichanneled river to where its wide alluvial mouth yawns to the Beaufort Sea.

We parallel the river along a well-trodden game trail that is covered with muskox tracks about the size of pony hooves. The round tracks are as wide as they are long. The front hooves are slightly larger than the back, since the bulk of the muskox's weight is in the head and shoulder region.

Many dwarfed shrubs are shrouded with *qiviut*, the soft, woolly hair of the muskox, prized by handspinners for its exceptionally high insulation value. Unlike the caribou, who have little underwool, relying on their hollow hairs for insulation, muskoxen have a tremendous coat of underwool, much denser and finer than sheep's wool. Muskoxen can afford to carry a heavier coat of fur since they are more sedentary than the migrating caribou, who must often outrun wolves.

"Hey, look at this stuff! There must be elephants walking around here," I say.

I walk around a huge mound of muskox dung that can hardly be referred to as droppings. The dung looks like small bales of weathered hay, with some piles measuring several inches high. I had read that muskoxen have a bulky high-fiber diet but never dreamed it would be this high.

A muskox has a rumen more than twice as large as the rumen of a caribou, and it is capable of ingesting large quantities of fibrous sedges and grasses. It has a wide mouth adapted for its bulk-feeding habit, while a caribou, a more selective browser, has a narrow mouth designed for eating easily digested plants such as lichens. Because the muskoxen and caribou have different feeding habits, they can be neighbors with minimal competition over food sources.

A little farther on we spot four muskoxen browsing along a large gravel bar in the middle of the Katakturuk River. At a distance of about two miles, they make an arctic grizzly bear look petite. Ave calls them "walking haystacks" because their massive, shaggy bodies lumber across the landscape on short, stumpy legs. The guard hairs of the muskox's dense furry skirt measure close to a yard long, easily draping below the animal's knees. The Inupiat simply call them *umingmak*, the bearded ones.

Through binoculars we watch the muskoxen browse on what looks like a species of *Oxytropis* and perhaps some dwarfed willow buds. Although it is June 19, there appear to be few available green plants on the tundra. Grasses are just beginning to send up their shoots, but the tundra is still predominantly tawny brown. Dwarfed willows have yet to leaf out, and there are only a few blooming plants, like the purple mountain saxifrages that hug the well-drained tundra. *Oxytropis*, which is high in protein, is one of few nutritious green plants available to the muskoxen. Once the willows leaf out, the muskoxen will thrive on them as their main summer forage.

In 1969 and 1970, a group of sixty-three muskoxen was shipped in wooden crates from Nunivak Island in the Bering Sea to the North Slope of Alaska. In an effort to reestablish a muskox population on the North Slope, fifty animals were released near Barter Island with the assistance of local Kaktovik residents. The remaining thirteen animals were released near the Kavik River, just west of the Arctic Refuge. At least ten muskoxen died from stress and the sedation necessitated by the transplant. A few disoriented animals wandered offshore onto the sea ice, and Kaktovik residents herded them back to land using snow machines.

One muskox wandered south over the crest of the Brooks Range toward Arctic Village. Kias Peter, an Arctic Village resident, was out hunting along the East Fork of the Chandalar River and was stunned when he saw the shaggy creature. He had never seen a muskox in his life, and he was unaware of the transplant effort. He ended up shooting the animal out of curiosity of how it might taste.

Muskoxen were scattered across the North Slope until the latter part of the nineteenth century. Also, there are a few reports of muskoxen on the south slopes of the Brooks Range. Although not a primary food source, historically they were hunted by Inupiat people, who utilized the meat and hides of the animal. Muskoxen are considered vulnerable to hunting, and it is believed that the introduction of firearms, coupled with an increased slaughter by Yankee whaling ship crews, led to their disappearance within Alaska.

In recent years the muskox population on the Arctic Refuge coastal plain has rebounded with a healthy 20 percent annual average growth rate. The total herd now numbers between 350 and 400 animals on the coastal plain, and some muskoxen have dispersed outside the boundaries of the refuge. They can commonly be spotted in small groups, browsing in the willow thickets along the many rivers and creeks that meander across the coastal plain. Occasionally groups as large as one hundred are seen.

As we continue toward the coast, we fight the steady wind, which has blown every drop of moisture out of my face. The surface of my cheeks feels like the frozen dried apricots in my windbreaker pocket. My shoulders ache from the frequent wind gusts that practically rip the pack off my back. During our trip we have experienced one day out of fourteen that has been windless; that day the temperature soared into the low 50s. It was the only liberating day when we could wear T-shirts and enjoy the brief sunshine. The rest of the time the temperatures have hovered in the mid-30s.

We hopelessly attempt to get out of the wind in the lee of a couple of big tussocks. Ave breaks down. During our stop he accepts a few of my last remaining m&m's. Last night he drank half of one of my hot chocolates. When Ave Thayer drinks or eats something with sugar, you know it's cold. You know he has burned up every last noodle.

Looking out across the windswept plain, I can see the tundra is flecked with the white spots of male rock ptarmigan. During the past week we have spotted scores of the visible males and frequently heard them deliver their comical snorting calls, proclaiming their territories. Unlike the females, who have molted and are well camouflaged, the males are pure white with the exception of the red combs that garnish their heads. The highly visible males help to distract predators, like the long-tailed jaegers, from their mates' nests.

Ptarmigan have apparently been prolific in this region for many years. During the winter of 1913-14, Vilhjalmur Stefansson and R.M. Anderson wintered near Simpson Cove with their Canadian Arctic Expedition and two of their three vessels. Stefansson and some members of his crew were stranded on shore after their ship, the *Karluk*, was caught in the sea ice and blown west in an arctic storm. The ship was later crushed by the ice and sank off the Siberian coast. Short on rations, the crew relied heavily on ptarmigan for winter survival, reportedly eating about twenty birds daily for two to three months. The region south of Simpson Cove and Camden Bay is still considered a good ptarmigan hunting area by Kaktovik residents.

"Things go on that are left alone," Ave said of the ptarmigan's continued presence in the area. Humans may have harvested many ptarmigan in this region, but they have not encroached upon the habitat.

Just before reaching the coast, we see our first sign of human activity during the two-week trip, a USFWS plane piloted by assistant refuge manager Don Ross. Don easily spots us walking along a gravel bar near the mouth of the Katakturuk and buzzes over our heads. He circles the single-engine Heliocourier, making a second low pass, and some small object flies out his window. We watch it land on the gravel bar within twenty-five yards of us, almost on target. We expect to find some important message in a can, but instead find a stale granola bar which has probably been tucked away in the plane for months. Still, it tastes good enough to devour as the plane putters off in the distance.

We gradually angle over to the coast through wetlands that remind me of rice paddies. Curving around numerous polygon-shaped tundra ponds, we spot pintail and old-squaw ducks, and many northern phalaropes. I pause to watch a group of female red-necked phalaropes probe for food in one of the ponds. Their slender black bills furiously needle, like sewing machines, into the muddy bottom of the ponds. They feast upon the benthic larvae as though they are desperate for calories.

Phalaropes are unusual in that the male and female breeding roles are reversed. Females sport the bright-colored plumage, while the males are duller and more camouflaged. The males are the nest builders and sole incubators of the eggs. Once the females have laid their eggs, they leave the nest site and play no parent role in terms of hatching or rearing their young. The male phalaropes become single parents, while the females often mate again with another partner, or congregate together, gorging themselves on insect larvae in preparation for their long southern migration. The small, lightweight phalaropes spend most of the year at sea, thriving on plankton and wandering to distant waters off the coast of South America.

As the female phalaropes repeatedly dunk their heads and bow their brick-red necks for food, I imagine these delicate seafaring birds bobbing at sea during a winter storm, thousands of miles from this tundra pond, at a latitude where Spanish is spoken. It all seems so impossible, so humbling. Tomorrow I will hop on a DC-3, sweep across the Arctic Refuge, and migrate to the grocery store in Fairbanks. So easy, so predictable. The only adversity I might face would be a plane delay or mold in my refrigerator. I think to myself that when I sit down to eat dinner tomorrow night, and during my hot soak in the tub, I will remember the probing bills of the red-necked phalaropes.

We soon reach the shores of Simpson Cove and find old signs of human presence. We walk by some driftwood house ruins that are quite a contrast to the DEW station's tin cup–shiny buildings in the distance. The house ruins are probably the remains of a trading post used by several white traders in the 1910s and 1920s. During that same time several Eskimo families lived on Nuvugaq (Collinson Point), across the cove from where we stand. Nuvugaq is still an important campsite and waterfowl hunting area for Kaktovik residents.

In my mind's eye I picture Simpson Cove over time. Inupiat driftwood and sod houses lining Nuvugaq; Captain Collinson stopping by in search of Sir John Franklin; white traders and whalers bartering with the Natives; Stefansson and Anderson wintering here; Eskimo families herding their reindeer; the DEW line men driving their weasel around the cove; and today's Kaktovik residents returning here to hunt and fish. The explorers, traders, whalers, and the DEW line men had all come and gone. Only the Inupiat continue to return here.

As we arc around the cove, I have mixed emotions about our trip, which is about to come to an end. On the one hand, I long for sunshine, a hot bath, and no wind. I can taste the fresh lettuce and green onions from my garden. On the other hand, I long to step through winter's window to watch the tundra burst with summer greenery, meadows fill with wildflowers, and chicks appear from their camouflaged nests. Mosquitoes by the millions would fill the air in just a few days, and with them tens of thousands of nestlings will wobble through the tussocks.

It was a trip to be remembered as almost summer. Yet I didn't feel shortchanged in the least. Over the years I have discovered that each minute spent in the Arctic—whether in a tent during foul weather, on top of a breathtaking glaciated peak, or in the midst of ten thousand caribou—carries the fullness of a rare wilderness experience. It is that fullness and the pure wildness of each waking moment that keeps bringing me back to the Arctic Refuge.

Part Two

INTO THE BROOKS RANGE

Bear track with my daughter's hand next to it

5

River with No Willows

OUTSIDE THE DC-3 WINDOW, endless peaks lift their majestic faces toward an ice-clear sky. Beneath the wing are summits upon summits, some 650 miles of interconnecting ridges and valleys, stretching westward across Alaska from the Yukon Territory to the Chukchi Sea. Looking back, I can barely see the northern edge of spruce forests. The distant trees are a faint shadow covering the valley floor of the East Fork of the Chandalar River. Well beyond timberline we cross the spine of America's northernmost mountains, the Brooks Range.

Along the Continental Divide, north-facing mountains are stroked by polar sunshine only a few months of the year. Shadowed cirques ring their craggy chins. Lonely steel-blue tarns eye surrounding mountains above the rubble of ancient moraines. Creeping glaciers finger their way toward scores of nameless velvet green valleys, curving around ridges and mountains like rivers winding from cutbank to cutbank.

The relief is impressively raw. No trees mask the landforms. Ice-mantled peaks flow onto windswept ridges, then onto naked, green valleys with shimmering waterfalls, rivulets, and streams gently pouring from the mountains. The valleys and sweeping tundra slopes soften the rugged mountains, the way the alpine meadows do in the high Sierra. There is great beauty in the pure nakedness of this land.

Far below, Dall sheep fleck the upper slopes of U-shaped valleys like pieces of white granite. We follow the winding Hulahula River, named by Hawaiian whalers who traveled past its mouth in the late 1890s. The braided river meanders to the Beaufort Sea, and perhaps reminded the homesick whalers of their culture's dance.

Soon we pass over the gentle northern foothills, and I see the white table of sea ice beyond the sweeping coastal plain about thirty miles away. I continually turn my head back and forth, glancing at the glaciated peaks, the treeless tundra plain, and the more distant polar ice. Since this is our first trip to the northern region of the Arctic Refuge, I'm in awe of the spectacular views in every direction; I'm like a child taking a first airplane ride.

Flying over this vast country, our plane is like a tiny insect lost above some huge grassy field. This airplane is an insignificant speck above the open, unobstructed coastal plain. We are a capsule of civilization peering down on this wildest corner of America.

Dennis and I, along with our friends Duncan Wanamaker and Alison Trembley, will soon backpack through a slice of this magnificent refuge. Our 150-mile route will take us from the mouth of the Okpilak River, across the coastal plain, to the Okpilak's headwaters in the Romanzof Mountains. Then we will cross glaciers that hug the Continental Divide, descend into the upper reaches of the Hulahula River and the East Fork of the Chandalar River, and hike south along the Chandalar's broad valley into the spruce forest. We will experience every link in the chain of arctic and subarctic ecosystems: from the treeless coastal plain and glaciated mountains to the boreal forest and exquisitely pure rivers that flow north or south from the crest of the Brooks Range.

"This trip is a chance of a lifetime. Look at all this wild country!" I say to wide-eyed Alison, one of Dennis's former students.

Dennis and I had just completed our first year of teaching at Arctic Village, about sixty miles south of where we would end our backpacking trip. As former teachers in California, we had spent most of our summers exploring wild places within the Sierra, Cascade, Olympic, Sawtooth, and Canadian Selkirk mountain ranges. We were hooked on the outdoors, and once calculated that while leading outdoor education trips through the school year and summer months, we had spent slightly more time sleeping in a tent than at home.

What I had thought was wild country in the "lower forty-eight" didn't begin to compare with what I was beginning to see of this roadless northeastern corner of Alaska, only slightly smaller than the state of Maine. No other place in America is as vast and wild. From the air we see ancient caribou trails etched into the tundra, other game trails following river corridors, and sheep trails zigzagging across limestone slopes. But here there are no man-made trails. Nor are there parking lots, visitor centers, concessions, outhouses, designated camping areas, garbage dumps, signs labeling natural features and mileages, nor the caution signs that we've all seen within other public parks and refuges.

Some of the more memorable signs that flash through my mind are: "Stay On Trail And Off Meadow," "Please Hold Railing While Ascending Lookout," "Don't Drink The Water," "Bear Area. No Camping." Then there are signs to protect people from people. Posted next to a ritzy hotel on Maui I saw: "Caution. Public

Beach Park. Enter At Your Own Risk." And, only in Alaska: "Caution. Children, Dog Teams, And Aircraft Use These Roads."

Within the Arctic Refuge we expect to meet nature on her wildest terms. Many of the valleys we'll explore, mountains we'll climb, creeks we'll cross, have no names. If we get lost, no signs will point the way during our month-long trek. Our maps, compass, and common sense will guide us. If a grizzly bear charges, we hope that blowing whistles, banging pans, and shooting flares will drive it away. As a last resort, we will also carry a rifle. Our only contact with civilization will be three food drops from a single-engine plane.

Several hours later Walt Audi shuttles us in his Super Cub from Barter Island to a small gravel bar on the Okpilak River, a few miles inland from the coast. The plane is overloaded, with Dennis and me shoehorned into the back and our backpacks strapped to the wing struts. Walt soon circles a gravel, bar near the west bank of the Okpilak that looks too short for a landing site.

"He's not going to land *there*, is he?" I yell, over the roar of the engine, into Dennis's ear.

From the air the potential landing site looks less than one hundred yards long. Walt is an experienced bush pilot, but I'm still nervous. If he overshoots the strip, we will crash into the river. Before I can contemplate the consequences, there's a tremendous thump, and we're on the ground, with rocks and a cloud of sand trailing behind, like in a roadrunner cartoon. Walt hits the brakes, and we come to a quick halt before rolling off the bar. The heavy load in the back of the plane keeps him from nosing over.

Warm temperatures and a haze of mosquitoes greet us as we unload the plane. Walt keeps the propeller whirring to blow the bugs downriver. Within a few minutes he's airborne again, leaving us in a cloud of sand and dust, with my hat sailing off into the Okpilak. Soon he'll return with Duncan and Alison. Once the putter of his engine fades away, we're in ecstasy despite the mosquitoes. At last civilization is hundreds of miles behind us. We are truly alone in this wildest of places.

What a view! It is July 17, and the coastal plain is peaking in summer greenery and wildflowers: fields of lavender lupine, white arctic milk vetch, yellow *Oxytropis* and paintbrush, pink bistorts and bumblebee flowers, and the delicate eight-petaled white dryas, a member of the rose family. The blend of bright tundra flowers is far beyond my expectations. The fields of lavender remind me of spring meadows of lupine in northern California.

Beyond the sweep of green and bloom are the lofty, glacial-mantled peaks of the Romanzof Mountains, as awesome from the ground as from the air: Mount Michelson, Mount Hubley, Mount Isto, Tugak Peak, Mount Waw. These exquisite mountains, rising to heights of more than nine thousand feet, are the highest peaks of the entire Brooks Range, and perhaps the least climbed mountains by virtue of their remoteness. They beckon us to ascend them, and we carry glacier-climbing equipment in the hopes of scaling Mount Michelson.

In the summer of 1826, when Sir John Franklin explored the arctic coast of what was then Russian America, he was impressed by the heights of these mountains, and gave them their name in honor of Count Nicholas Romanzof, once chancellor of the Russian empire. Romanzof, like Franklin, was dedicated to promoting science and discovery.

We can't be sure where Franklin delineated the exact boundaries of the Romanzof Mountains, but roughly they cover an area sixty-five miles long, between the Hulahula and Kongakut drainages, and about twenty miles wide, between the northern foothills and the Continental Divide. The most prominent peaks within this region occur in an area about fifteen miles in diameter, between the Hulahula and Jago rivers. The Okpilak River cuts through the middle of this uplifted zone, one of the main reasons we had chosen it as our route of travel.

To the north, we look across a sea of glaring white ice stretching toward the North Pole. On this warm July day, strange mirages bubble above the ice in the distance. How odd to see arctic mesas and buttes wavering above this frozen desert, yet it reminds me that these high latitudes receive about the same amount of precipitation as the Mojave Desert, about six inches annually.

Out on the distant, drifting pack ice, thousands of polar bears roam across their homeland, stalking seals beneath the distant pale blue. Although we probably won't see a polar bear on this trip, it's exciting just to know the magnificent animals are there. In a few months female polar bears might be standing right here, looking for onshore denning sites where they can bear their young.

We set up our tent within a few yards of the Okpilak, meaning "river with no willows" in Inupiaq. The Okpilak has several braided channels, the main one more than knee-deep and quite swift. Although there is an array of dwarfed blooming plants on the river bars, indeed, there are no visible willows.

The air is dead calm as we drive in tent stakes, except for the thousands of mosquitoes that swarm about our heads. Although our skin is smothered with mosquito repellent, the pests still form black clouds around our faces, and their incessant high-pitched whine is aggravating. Every so often one of us inhales a couple through the mouth. There's the distinct cough-gagging sound, followed by forced exhales, spitting, and a curse of sorts. They are worse than what we imagined.

Just before erecting one tent Dennis notices that the door is wide open and the inside is already thick with mosquitoes. He quickly zips the door, then rolls over the flattened tent, swatting the nylon with his hands. As if this pathetic combat will do any good; no sooner will we crawl into the tent than the swarms will be right on top of us. We put on headnets for a while to avoid swallowing the pests.

Arctic ground squirrels chatter at us from a small pingo that is full of holes from their tunneling. These hefty squirrels appear fat compared to the squirrels I saw several weeks ago near Arctic Village. They have obviously packed on a month's worth of summer forage. Curious creatures, they bounce around on their hind legs, scurry several yards toward us, then run back to their homes.

Within an hour of our arrival, we spot a fox prowling near camp. The squirrels immediately disappear into their pingo once the fox is in sight. The tawny brown fox, with black hind and front quarters, trots up to the ground squirrel den, pokes its snout into a few of the holes, then moves on after making the passing threat. It lopes along, sniffing the tundra as it circles our camp, perhaps looking for rodents or a nest of chicks.

The fox may soon find a meal. Nesting birds surround our camp: ptarmigan with their striking red crescent brows, soaring long-tailed jaegers, gulls galore, plovers and sandpipers, and the ever-present Lapland longspurs. We hear them constantly as well as see them, marveling at the variety of birdcalls as much as at the intensity of activity.

Later we watch a mountain storm brewing over the Romanzof Mountains. Junco-gray clouds gradually swallow some of the glaciated peaks. Lightning bolts illuminate the dark clouds and shadowed ridges, but we are too far away to hear the rumble of thunder.

"Earth and the great weather have carried me away and move my inward parts with joy," Dennis says, quoting from the Eskimo passage by Osarqaq from Kenneth Brower's classic book *Earth and the Great Weather*. We had soaked up Brower's book, poured through the exquisite pictures many times, dreaming of the Brooks Range. Now those abstract images, those dreams, had become very real.

❖

Two days later we are ten miles closer to the Romanzof Mountains, and with each stop we take the peaks grow ever more impressive. The spectacular view of the north face of Mount Michelson takes our minds off backpacking discomforts. We have yet to break in and become accustomed to the sixty-pound backpacks, the hordes of mosquitoes and the slogging through marshy tussocks. My calves are killing me from sinking into the soggy ground and lifting my iron-heavy boots over foot-high tussocks. This is the toughest walking ever.

Coastal plain with Mount Michelson in background

New to the coastal plain, we make mistakes. We hike in the middle of the day in the sweltering heat, with the temperatures soaring into the high 70s and low 80s. I sweat off mosquito repellent as fast as I put it on. Instead of following the game trails of wiser animals, along the winding banks of the Okpilak, we decide to take shortcuts. Dennis and Duncan are avid map readers, and of the firm belief that the shortest distance between two points is a straight line. Every so often they pull out their maps and make futile attempts to figure out exactly where we are in relation to what bend in the river.

"If we bear due south here, we'll avoid making this west-east detour along this big bend in the river," Duncan suggests, pointing to his map. Dennis agrees. Alison and I are tempted to stick with the firm walking along the river, but we stride off through the marshy tussock fields anyway. We all learn the hard way, squishing our way through the swampy terrain, tripping over the grassy mounds.

After trying a few of those shortcuts, we soon realize that the extra energy expenditure of plodding through the swampy terrain exhausts us. We may save a mile or so in walking, but the shortcut is a killer on our bodies. Our best bet is to follow established animal trails along the Okpilak, or to walk the gravel bars and old floodplains, where the ground is firm.

Along the silted Okpilak I try imagining what the river's headwaters look like. Our topo map shows two huge arms of glacial ice hugging a massive rib of mountains between the east and west forks of the upper Okpilak. The hot temperatures are obviously causing these glaciers to melt at a rapid rate, sending ancient yet fresh water to mingle with the Beaufort Sea. As this turbid torrent rushes by me, I can visualize centuries of winter snowfall transformed into ice, now released from time on its liquid journey to the sea. Yet soon this water of glacial origin will once again turn to a mantle of sea ice. Snow to ice, to water, to ice; each form of the same element molding the landscape over time.

Just before reaching our next campsite, feeling dead tired, I see something that changes my perspective on weariness. Across the river, slumped on an old snow patch is a traveler who has trekked much farther than we. There, on the cooling drift of snow, is a feeble, fly-ridden cow caribou, collapsed and dying. The poor animal can barely lift her head as botflies encase her nostrils and eyes, and warble flies burrow into her body. Her ribs protrude through an ashen, mangy coat of fur, and I wonder if some predator will put the caribou out of her misery. Perhaps the cow is crippled or suffering from some disease or parasites. What is this caribou thinking as she faces her lonely death? Did she bear or lose a calf this year? And who would know or care of the thousands of miles she has traveled?

I continue another mile, thinking of the dying cow, while following a well-beaten caribou trail. Every square inch of my path is covered with overlapping cow and calf tracks. I feel somewhat humbled hiking this relatively short distance compared to the thousands of miles these caribou travel each year. The great herds migrated eastward a few weeks ago. All that remains in their wake are the old tracks, scattered antlers, and the lone, dying cow.

The image of the fly-covered caribou lingers. In addition to fighting the harassing mosquitoes, caribou must contend with warble flies and botflies that emerge in July and August. The warble fly lays its eggs in the caribou's fur, usually in the stomach area, and the hatching larvae penetrate the skin and migrate through the intestinal cavity to the caribou's back. There they grow by the hundreds on the underside of the skin through the fall and winter. By spring the larvae measure about a half-inch long in their watery capsules. In early summer the warbles cut breathing holes through the caribou's skin, then exit their host to complete their growing cycle. Summer caribou skins are infested with tiny holes where the warbles have emerged, and are therefore undesirable for Native peoples, who use hides for clothing and other needs.

I picture Margaret Tritt in her home in Arctic Village teaching me how to flesh out a winter caribou skin with her scraping tool handcrafted from a moose leg bone. As I learned to peel away the fleshy part of the skin with a steady hammering motion, I frequently hit warble fly larvae that bubbled under the skin. In some cases the juicy larvae punctured, sending a spray of liquid into my face.

Whereas warbles attack the caribou in the abdomen, the botflies plant themselves in the nasal passages. Caribou reportedly detest the botfly and will run away from them when attacked. The bot larvae grow inside the nasal passages and throat, and this can affect the caribou's breathing. When caribou are heard coughing in late spring and early summer, it is because their throats are aggravated by the parasite.

Mosquitoes, parasites, and predators. Thousands of miles of annual migrations. Thousands of days of cratering through snow to obtain food. Life isn't easy for these wandering grazers. Yet because they have evolved in high latitudes where few people have settled, their far-ranging habitat has remained relatively free from human encroachment. Many species dwelling in more temperate zones have simply lost their habitat to humans. Some adapt, others grow rare and endangered, and some species become extinct. Although the life of a caribou appears to be a struggle in the Arctic, they at least have had the increasingly rare gift of wild land to sustain their free-roaming species. But for how long?

In the evening we spend much time watching a pair of lesser golden plovers who built their nest on a gravel bar near our tent site. The ground nest of sticks and grass looks vulnerable on the open river bar, yet it blends well with the landscape. After a time the female leaves the nest to feed with her mate. Golden plovers are striking in their summer plumage, with a velvet black belly and neck, and tiny golden nuggets flecking their backs and wings. They stand proudly, as if knowing they're cloaked in beauty.

While the pair bob their heads and pick insects from the river bar, we walk over to their nest to take a quick peek at three creamy white eggs blotched with brown. No sooner do we approach the nest than the pair immediately return to protect their territory. The female hurriedly lands near the nest, then scurries away while pointing her tail feathers downward, squeaking at us on the run. The male begins his broken-wing display, also trying to distract us. He hobbles about, flapping and dragging one wing on the ground, as if injured. We leave the site quickly, as we mean no harm, yet we appreciate the opportunity to witness the birds' defensive behavior and gain a better understanding of how they protect their offspring on these open river bars.

In less than a month these long-distance marathoners will flock together en route to their wintering grounds in South America. I picture the planet from above and imagine what their 20,000-mile migration loop might look like: east along the coast of Alaska across the Yukon and Northwest Territories, over the northern straits of Hudson Bay, across northern Quebec, along the coast of Labrador, then south beyond the Gulf of Saint Lawrence and Nova Scotia, over the open Atlantic, past the West Indies, on to the coast of Brazil, and then to another coastal plain. The golden plovers reach their final destination: the Pampas of Argentina, the grassy plain region that rolls east from the Andes to the Atlantic. Coming home to the Arctic, most plovers take a shorter route—through the interior United States, up the Mississippi Valley, and on through central Canada to the northern coast. I'd give anything to pitch a tent, six months from now, out on the Pampas next to these same two birds.

Before falling asleep I gaze out the tent door to the glacial-mantled summit of Mount Michelson. In the late evening light Michelson's rose-tinted glacier looks less sheer and more like a gentle sweeping bowl that bridges the summit to a surrounding ring of smaller peaks and ridges. Step by step we have gained a closer perspective of this exquisite mountain.

Back at the coast we had stood at a distance from the Romanzof Mountains and gained appreciation of the wholeness of the entire range; it was much like viewing a prized painting from across a room. As we've moved closer, subtle details of the mountain relief have begun to stand alone. We study the intricacies of individual mountains, glaciers, and ridges highlighted in the glow of midnight sun. The distant, wide-angle, and close-up views enable the mind to capture the fullness and special features of the scene before us.

I think of other mountains visited and the value of the open meadows, the grassy plains, and the sagebrush sweep of land that borders them: the east slope of the Sierra, the meadows beyond the Grand Tetons, the Sonoran Desert at the base of the Catalina Mountains, the prairie stretching eastward from the Rockies. It is the open roll of these lowlands that frees the eyes, mind, and soul and allows us to appreciate the dramatic rise of the highlands. What are the mountains without the plain?

A few days later we hike along the Okpilak in midnight light. Hot temperatures have persisted, so we decide to backpack in the coolness of evening, when the colors are richer. We walk several hundred yards from one another so we can fully absorb the views and sounds without having to trail behind the obtrusiveness of someone's overstuffed pack.

This is the most enjoyable time to backpack across the plain: in the coolness, along the shimmering Okpilak, with the fiery sun rolling just above the distant pack ice behind us. Sunset colors spill across the northern horizon, turning fields of lupines iridescent, dried mosses golden, and the turbid river a bubbling rose beneath salmon-pink clouds and sky.

Looking down the Okpilak I can see the entire course of the river, its braided rivers of sunset meandering to the sea. We have gradually gained about six hundred feet in elevation and can now look back along the forty-some miles we have traveled. To the southwest rises Mount Chamberlin, its glaciered silt towering above lakes Schrader and Peters, just west of the Hulahula drainage in the Franklin Mountains. The arctic sunlight filters through broken clouds, painting Chamberlin's cornice a rose lavender.

A few miles farther on I see Dennis in the distance waving his arms with his pack off, a signal that he has found the next campsite. He has reached Okpilak Lake! This is the imaginary lake we have all dreamed of. A lake that marks the gateway to the Romanzof Mountains, and the end of our tedious journey across the coastal plain. A lake that we can swim in and eat fish from. A lake that should mirror the majestic Mount Michelson. A surge of energy sweeps through my sore calves and feet, and I pick up my snaillike pace.

Within a few minutes I hike over a gentle slope through the forever tussocks, and there is Okpilak Lake nestled beneath the shoulder of the Romanzof Mountains, with Mount Michelson's glaciers curving downward toward us on the other side of the river. Sunrise colors paint the exquisite lake a deep lavender, and the joyous song of the Lapland longspurs and the piping sandpipers welcome us. All I want to do is collapse and breathe in the surroundings. Dennis still has enough energy to fish.

"Hey!" he yells. "I've got one on." His rod bends in a graceful arc toward the fighting fish, and within a few moments he lands an average-size arctic grayling on the bank. Its transparent silver-blue dorsal fin is illuminated, like backlit lupine.

Before long he has caught another grayling, and we cook the fish over our first open fire of willows—now that we have hiked into the foothills, we are beginning to see a fair amount of willow shrubs surrounding the lake. We have a delicious meal which we appropriately call dinfast. Ravenously we eat the lemon-peppered fish and rice and shortly after fall asleep to the soft piping of sandpipers in the twilight that separates our days.

We awaken to a desert-hot sun, and our tent is like a furnace. The thermometer in the sun reads eighty-six degrees. Unknowingly I slept with my hand resting against the mosquito netting and am surprised to find that dozens of mosquitoes have turned my fingers and hand into a pegboard of bites. I wonder how much blood they needled out of me. Dead asleep, I missed their feasting.

We crawl out of our tents like lethargic lizards to greet the sun. I would never have guessed it could be this hot on the North Slope of Alaska. I strip off my clothing and plunge into the lake, which is warmer than anticipated. Diving beneath the surface, I try to lose the mosquitoes that have tracked me into the water. When I surface, there isn't a bug in sight, and I wish we had brought an inflatable raft.

Dall sheep fleck a steep talus slope that faces Mount Michelson. From their vantage point, what a view they must have, looking across the valley at lofty peaks and glaciers, and beyond this lake, across the coastal plain. I wonder if they see the commotion of a floundering human in the lake.

Out of the water I dry off and dress quickly to avoid the swarms of mosquitoes. A spotted sandpiper lands in front of me for a few moments, then takes to air again. This small sandpiper is an entertaining, curious bird. He darts about in the wind like a swallow and frequently plunges into the lake. Two white stripes decorate his air-streamed wings, and his white breast is noticeably speckled with grays and blacks. His coral black-tipped bill is striking, as it merges with an ivory white stripe that arcs above the eye. The sandpiper stayed around our campsite during most of our first relaxing layover day.

In the late afternoon I walk up a knoll to gain just enough elevation to see the Okpilak River wind its way to the Beaufort Sea. Looking back at the country through which we had just hiked, I have no way of knowing that in the future, winter seismic trains will explore for oil and gas, and crisscross our route of travel. I have no way of knowing the Department of Interior will propose to Congress that this precious coastal plain be opened for oil and gas development. I cannot know that a future oil development plan will suggest construction of numerous drilling pads, a permanent airfield, and spur roads and pipelines crossing the Okpilak River just ten to fifteen miles downriver from where I now stand.

Whether Congress will accept such a short-sighted development scheme here, beneath the shoulders of Mount Michelson, under the amber eyes of onlooking Dall sheep, in the wildest corner of America, is an unknown.

6

Through the Romanzof Mountains

WE LEAVE OKPILAK LAKE in the splendor of late evening colors, our bodies and minds revived and prepared for mountain travel. Within a mile of camp, the mouth of the valley begins to close upon us, giving us the sense of being funneled from the open coastal plain into the throat of the mountains. The gentle foothills surrounding Okpilak Lake and the expanse of tundra are now behind us. With each step forward, we see new rugged escarpments emerge above sharply rising valley walls and steep talus slopes. Mount Michelson's hanging glacier boldly rises above the valley, cutting into a rose-drenched sky.

We follow our elongated shadows as the sun makes its subtle roll across the northern horizon. The low-angling light creates rich tones of contrasting colors. Deep gray talus slopes tumble onto the verdant V-shaped floor of the valley. Copper-brown rock slides swirl around charcoal mountains. Tangerine and yellow lichens gnaw at kettle-black rocks that are scattered near the caribou trail we follow. Illuminated dwarfed wildflowers, such as the mountain avens and pink phlox, are brightly clumped along the banks of the Okpilak.

None of us wears a watch, but we've learned to judge the time of day by the sun's rays. In the late evening the east wall of the valley softly glows. Pale gray mountains turn to silver, and tundra slopes turn a shamrock green. As the golden orb rolls northeast above the pack ice, it gradually reaches the center of the valley, sending a flood of soft filtered light toward us at the lowest possible angle, an angle like that of a plane in relation to the ground just before touchdown. It is close to midnight. Soon Mount Michelson and the west side of the valley gradually become candlelit in the soft pink of dawn, and we celebrate morning. During the late evening and early morning hours, this mile-wide valley is gloriously alive with a never-ending light show contrasted against the immense shadows of towering mountains. Such shadows temporarily swallow the landscape and are our only hint of night.

Mount Michelson's hanging glacier is impressively closer now. Much of its sparkling ice-blue tongue is snow covered at the upper elevations. Above the rubble of its old moraine, the exposed icy snout looms some one thousand to fifteen hundred feet above us. My skin feels the cold breath of the glacier in the evening's down-valley breeze, and we can hear the countless rivulets of melting ice invisibly trickling down the glacier's rock-cluttered trail toward the Okpilak.

As we gain a few more miles, the valley walls gradually palm together. The talus slopes and underlying lateral moraines grow more sheer. I pass one mountain on the east side of the valley whose face has been violently uplifted. One thick band of folded sedimentary rock tells the story. Half of the band runs horizontally across the mountain. The other half takes a sudden right-angle turn to the sky. A perfect backward L.

The narrower the valley becomes, the louder the Okpilak's roar. The river is a rage of silted glacial water with rumbling boulders spilling down from its tributaries. Looking at the map, I count at least sixteen tributary valleys that are fed by glacial fingers that grip the summits of Mount Michelson, Tugak Peak, Mount Isto, and numerous nameless peaks near the headwaters of the Okpilak. With the warm temperatures the river is a flood of glacial icemelt. How would we ever cross it to reach Mount Michelson?

From the beginning of our trip, we hoped that the Okpilak would recede enough for us to reach its west bank. Yet with the continual warm weather and high water, it seems unlikely that fording the river will be possible. We set up camp near a nameless tributary that offers an uplifting view of Mount Michelson's glacier. The unreachable glacier is only about three miles away, and all of us are disappointed in knowing there is little chance of our scaling Mount Michelson. Looking out the doors of our tents, we can only drool at the exquisite glacier that drapes the mountain, and dream of its hidden summit. Also, there are inviting hot springs on the other side of the river.

The next day we decide to explore the tributary valley that would lead us to Fox Peak, a 7,610-foot mountain whose north face wears a pair of small remnant glaciers. Day hiking without the heavy backpack makes me feel weightless, as though there are springs attached to my boots. There is great joy and liberation in being able to walk freely, stop, pause, and study the rugged ridges that now envelop us, without the burden of heavy packs.

With each few feet gained in elevation there are fewer mosquitoes, and we pick up a cooling mountain breeze off the heels of unseen glaciers. Soon we are in our T-shirts, with hats off and no need of mosquito repellent. Our skin breathes for the first time without being smothered in bug dope. All of us are in good spirits as we climb over granite boulders amidst the jumble of sedimentary rocks, sandstone, and shale.

Until today we heard only the dominant roar of the turbid, swift-flowing Okpilak River. As that river fades in the distance, the gentle sounds of this high-mountain stream ease us up the valley. Delicate trickles of crystal water lace the rocks beneath our feet, gurgling and tinkling through the channel. It is as though we have left the din of a loud brass band, and now can hear the soft melodies of individual woodwind and percussion instruments.

Gaining more elevation, we gradually become hemmed in by steep canyon walls. Our once-braided stream turns into one or two distinct channels, and we find ourselves wading through the water, skirting around sparkling cascades, and being careful not to slip on moss-covered rocks. Soon we reach a thunderous waterfall that terminates our ascent up the valley. Surrounded by a mass of twenty- to thirty-foot interlocking walls, there is no way to free-climb our way above the falls.

"No way to get above this without ropes and pitons," Dennis says.

I spot one small moss-matted ledge near the falls and scramble up the rocks to get a perching view of the chain of bubbling pools below me. The ledge is buttressed by angular blocks of rocks that wedge and lean into one another. They form a sturdy wall, as though mortared together. Most of the rocks are covered with timeless circular patterns of lichens, etched upon the stonelike cave paintings.

The predominant lichen is bright orange, the color of puffin bills. I place my fingers near the rootlike strands of the creeping lichens and feel rock particles the size of bread crumbs. The twisting feet of these plants, known as rhizines, are chiseling away at the rock, turning it to soil over time with the help of acids secreted by the lichens. As winter sets in, water that has seeped into the lichen-fractured rock will freeze and refreeze, resulting in further erosion.

Rock lichens are one of nature's most subtle and ancient erosional forces, breaking down mountains over centuries. There are hundreds of different species of lichens in northern Alaska, including many species of rock lichens. Research has documented that some parent lichen plants in northern Canada and Greenland have grown to be two hundred years old. These tenacious lichens slowly turn stone to sand, leaving scattered pockets of grated rock where other new plants can root themselves.

Hardy lichens are extremely resistant to drought conditions and cold temperatures. Lichens have been discovered even inside rocks in Antarctica. While vascular plants lose their leaves or die off in the face of winter, lichens carry the same appearance whether beneath a layer of snow or during summer months on the tundra. Yet, like other arctic perennials, lichens shift into a winter dormant state. Once the temperature climbs above freezing, the spongelike lichens become opportunistic, absorbing whatever moisture and nutrients are available directly into the plant's gelatinous tissue. Lichens avoid the complication of having to gather their food through a root system.

Interspersed between the lichen-encrusted rocks is an array of wildflowers that have found their niche, many rooted in lichen-produced soil. Bright yellow cinquefoils, lavender phlox, purple saxifrages, and forest-green heather are thriving on the constant spray from the falls. Within this productive grotto, nestled within a fractured rock, is what looks like a maidenhair fern. I expected to see plants such as the hardy, perennial saxifrage in these rocky crevices, but a fern? What a welcome surprise.

There is something about delicate ferns growing within circumpolar regions that seems miraculous. Perhaps, mistakenly, I've always viewed ferns as one of nature's more fragile, translucent plants. Growing up in the San Francisco Bay Area, I was accustomed to seeing ferns in soft-filtered light, their fronds illuminated like stained glass beneath the sheltering canopy of giant redwoods and Douglas firs. John Muir, while hiking in the Sierra, once eloquently spoke of fern fronds as a "living ceiling" that revealed "arching branching ribs and veins of the fronds as the framework of countless panes of pale green and yellow plant-glass nicely fitted together."

These translucent ferns of the temperate zones have, like lichens, evolved to be one of the most ancient, successfully adapted plants. Aerodynamic fern spores have blown to all corners of the world. Where there is enough moisture, some warmth, and soil, a fern will appear: near a shaded pool in the Sonoran Desert, within Kilauea's crater on the island of Hawaii, or in a rock's crack on the north side of the Brooks Range. These gentle yet bold ferns can be cradled by giant redwoods, but they can also flourish within deserts, volcanoes, and in the Arctic.

The surrounding mountains and jumble of rocks below me have the same timeless feeling of lichens amid the boldness of ferns. If we could have visited this region about 300 to 350 million years ago, there would have been no mountains, although the granite core that is now the heart of the Romanzof Mountains would have been thousands of feet beneath our feet. Instead of walking through mountains and valleys, we would have spent our time wading or snorkeling through shallow salt water, looking at live corals, and roaming beaches nearby. It is theorized that this one-time mountainless region, extending to the south side of today's Brooks Range, lay near the edge of the North American continent. The southern part of Alaska had not yet merged with this so-called Arctic Alaska block.

Opposite: *Brooks Range, aerial of Romanzof Mountains*

Then by about 135 million years ago, a major collision began the process that would gradually shape Alaska to be as we know it today. The North Pacific plate, carrying volcanic islands such as today's Aleutians, collided with the more stationary Arctic Alaska block. Over time mountain ranges such as the Brooks Range and Alaska Range were squeezed upward from the tectonic pressure, and the seawater receded. Beginning about 60 million years ago, the northeastern Brooks Range began to rise.

The Romanzof Mountains are unique in that they have been uplifted to become the highest range of peaks in the entire Brooks Range. What makes the Romanzofs even more interesting is that much of the range's massive buildup has occurred over a relatively short time, in the last half million to two million years. Glaciers of the late Pleistocene epoch that covered the Okpilak and many of its tributary valleys had little time to gnaw and polish the bedrock granite in any one particular zone. The mountains continued to rise quickly during the ice ages.

The Romanzofs show old wrinkles of the past in their conglomeration of limestone with fossilized coral and the wide variety of other marine life, sandstone, shale, and volcanic rocks. The recent uplifting has exposed a layered cake of rock formations that has baked over the past one billon years. On the northern front of the Romanzofs, uplifting has exposed the top layers of the cake—younger rock deposits, some 150 to 360 million years old. Intruding through the middle of the cake has been a tremendous batholith of granite, rising dramatically above the fossils and ocean-rippling shale. As one geologist said to me, just about every formation has been exposed in the Romanzofs by the series of uplifts.

Living amidst the rock museum of old and new geologic forces is perhaps what is most impressive about the Romanzofs. Here you can witness the birth of granite mountains. You can watch the ever-rising granite peaks working against the erosional forces of past and active glaciers and of rivers that cut through the heart of the mountains and old lateral moraines. You can see the weaker, softer rock peel away and tumble into the Okpilak. Looking down this tributary valley, you see scattered pinnacles barely holding their own, like old, crumbling buildings surrounded by skyscrapers. Each year more of the outcroppings are chiseled away by snow, ice, wind, and the subtle lichens, their debris rolling down the mountainside into the streambed.

Sometimes I wish I could watch mountains such as these grow the way I can watch the annual spring growth of birch and aspen leaves. If only this range could visibly unfold its millennia of geologic stories so that the eye and mind could comprehend the magnitude and the sweep of our planet's forces rather than piecing together the enormous puzzle of stones; then perhaps we would become humbled, less anthropocentric. Perhaps we would curb our insatiable appetite to conquer and consume the entire earth. Wouldn't such a vision make most of us stagger in our insignificance?

My thoughts turn to Supreme Court Justice William O. Douglas, who visited the Arctic Refuge during the late 1950s with Olaus and Margaret Murie. Douglas, a strong proponent of wilderness, became an advocate for protecting the northeastern corner of Alaska as a wildlife sanctuary. As a young teenager, I was not familiar with Douglas's association with the Arctic National Wildlife Range, yet his writings on other wilderness and social issues deeply impressed me. While in high school during the Vietnam years, I wrote down one of Douglas's passages and framed it with some autumn leaves; it hangs in our home in Fairbanks. It reads: "When man pits himself against the mountain, he taps inner springs of his strength. He comes to know himself. If man could only get to know the mountains better, and let them become a part of him, he would lose much of his aggression."

Descending the valley, I am invited by the warm sun to dip into an ice-cold bubbling pool beneath a series of small cascades. The water's temperature is no warmer than that of a drift of snow. I bolt out into the sunshine within seconds, gasping for breath. With no mosquitoes I find a smooth block of warm granite to rest upon. My skin tingles in arctic warmth as each bead of water evaporates. Soon the rock grows cold, and I move to another scoured boulder of granite. It is hard to imagine the glacial force that must have moved this boulder that is four to six feet in diameter. Would another glacial age send it farther, perhaps rolling it out upon the coastal plain?

Walking down the valley, I pause just above the point where the cone of talus pours across the green floor of the Okpilak's valley. From here I can see up the valley, some eight or nine miles south, to where the river forks around a massive ridge of mountains. The valley continues to narrow into what looks like a half-mile-wide trough. The river cuts deeper, and deeper, through the layers of talus and glacial deposits. Within a few miles it disappears entirely, becoming a hidden seam on the floor of this rock-filled trough.

Most of the valley's floor is covered with huge talus cones and alluvial fans, spilling into one another from both sides of the valley. The enormous fans of rock debris look like split-open sacks of rice pouring into one another. It is almost as though the pressurized core of granite mountains is heaving upward and bulging inward toward the valley, sending mountains of old rock debris tumbling down its tributaries, and squeezing the lateral moraines and talus into one another. In reality, it is gravity and the incising force of water spilling from the glaciers and ice caps that have pushed this rock debris to the valley's floor.

Cranking my head skyward, I find the vertical relief of the Okpilak's valley impressive. Standing at nearly 2,000 feet above sea level the glaciated shoulder of the 8,855-foot Mount Michelson looms above me only a few miles away. The vertical rise of the valley's slopes is so steep that it's impossible to see the summits of several nearby 8,000- to 9,000-foot ice-capped peaks. Looking at neighboring valleys to the east and west of the Okpilak, I see no spot where topographic contours are more squished together in so short a distance.

When geologist Ernest Leffingwell explored the Okpilak River valley during the 1910s, he noted that the Okpilak showed the greatest signs of glaciation of the several North Slope valleys he had explored. He estimated that the Okpilak's glacier deeply gouged a forty-mile-long trough through granite. He noted that at the river's fork two massive glaciers at one time merged together, filling the valley with three thousand feet of ice. Other geologists have noted that some of the glacial signs at the higher elevations could be part of an older, glacially carved valley that has been recently uplifted. Whatever the case may be, how impressive to imagine this valley filled with such a massive tongue of ice!

Over the centuries, the Okpilak River valley has gradually filled with the great alluvial fans and talus cones, while the Okpilak's east- and west-fork glaciers have receded. The slopes of rock debris and buried lateral moraines have somewhat hidden the U-shaped nature of the valley. If we could remove this veneer of talus and thick morainal deposits, and thus reveal the effects of glacial scouring and polishing, the Okpilak might look similar to a mini–Yosemite Valley.

Later in the evening we backpack about three miles to a tributary stream fed by the hidden glacial arm of Mount Isto. The 9,060-foot Mount Isto is the highest peak of the entire Brooks Range. It was named in honor of Reynold Isto, a USGS civil engineer who mapped much of the Brooks Range in the mid-1950s. Although Mount Isto is only five or six miles away as a raven flies, our view of the ice-capped peak is unfortunately blocked by the valley walls that hem us in.

The walk to what we call Isto Valley is spectacular in the glowing midnight light. The cumulus clouds have vanished, the sky is deep sapphire blue, and the turbid Okpilak still roars. We walk on a narrow carpet of tundra, wedged between steep talus slopes and the curving, heightening bluffs along the river. Gradually the valley walls palm closer together, and the band of blue above us grows more narrow.

The well-drained alpine tundra is quite different from the swampy, tussock-pecked tundra we had squished through on the coastal plain. Walking on the firm, relatively level tundra is a joy. Without the worry of tripping, we can look around and study the new vegetative zone.

Most of the alpine tundra's perennial greenery is made up of dwarfed birch and creeping willows. A new array of mountain flowers is interspersed between the crawling evergreens. Arctic bell heather, reminiscent of the Sierra's white heather, so deeply loved by John Muir, laces the tundra with thousands of tiny white bells growing only a few inches above the earth. Bright pink cushions of moss campion and plumes of bistort are sprinkled amidst the white bells and evergreens. Every so often we pass a bright bush of yellow cinquefoil, or the small sunflower-bright blossoms of the thyme-leaved saxifrage. A biologist once told me that exploring tundra vegetation is just like walking through a jungle, only this jungle is a foot high.

Before long we reach Isto Valley and the thunderous roar of its glacial-fed river. Boulders roll and rumble down the floor of the river while gravel grates its way to the Okpilak beneath the trembling din. The erosional forces are at work here. We wonder whether it will be possible to ford this raging tributary. This is the kind of surge that transforms angular rocks into rounded boulders, and our legs into frozen, feeble sticks. We decide to make camp and wait for the waters to recede.

The next morning the water level is down enough for us to cross safely. The rumbling boulders have toned down, although we can still hear rocks slowly inching their way down the floor of the tributary. We decide to cache most of our supplies and explore the upper reaches of the Okpilak with light packs. We take two days of rations and our glacial-climbing gear, then set out toward the headwaters of the Okpilak. We plan to climb the east fork's glacier to a pass that will lead us into the Hulahula River drainage. If we can't follow that route, we'll have to backtrack.

It takes several minutes to find a place safe enough to cross the ten-yard-wide tributary. Our feet turn to ice cubes as we carefully pick our way around boulders through knee-deep white water. Halfway across the stream, I begin to feel dizzy from staring at the rush of white foam. Each time I plant my boot on the floor of the stream, I can barely feel whether I'm on a firm streambed or moving boulders. When I reach the other side, I notice a small amount of blood on my leg. A boulder had grazed me, although I didn't feel it.

Farther up the valley we become more and more enveloped by steep limestone and granite walls. We tediously cross talus fans that pour into one another from the east and west tributary valleys. The river has gradually turned into one serpentine channel, cutting through years of rock debris. The more elevation we gain, the deeper the Okpilak's gorge becomes, and the louder the sound of the turbid river. Sheep and caribou trails are etched into the talus slopes above us, although none of us spots any large mammals today. We do hear the occasional whistle of a pika.

"Wolf tracks!" Dennis says, pointing ahead. All of us have yet to see a wolf in the wild, so we're excited knowing that at least one wolf traveled past here. The large set of tracks looks fairly old.

After a few hours of picking our way through lichen-scabbed talus, we reach an open green plateau near the Y of the Okpilak's deepening gorge. The river's roar is muffled in the chasm some thirty to forty feet below us. There is no sign of the entrenched river unless we peer over the lip of the plateau, down the river's canyon walls.

For the first time during our trip, the sky looks threatening, with thick slate clouds. We decide to set up camp on the inviting level ground before it starts to rain. Our smooth tent site is surrounded by white dryas and heather, and embraced by rugged ridges and steep valley walls that rise some four thousand feet above us. Our tent door offers a spectacular view of the sloping flower-specked plateau, the sudden drop-off into the Okpilak's chasm, and beyond, to the east and west glacier-gouged forks of the Okpilak.

On the west side of the river, just below the Y, is one of man's strange footprints. An old rusted boiler sits on the tundra, looking out of place. It is reminiscent of some of the discarded oil drums we've spotted along the pristine arctic coast. Later we would learn that the boiler belonged to a Norwegian trader and prospector, John Olsen. Olsen operated a couple of trading posts on the Beaufort Sea coast during the 1930s. In the springtime Olsen searched for gold along the Okpilak River and used the boiler to help thaw out frozen soil.

It begins to drizzle. We have turned into arctic desert rats, so the rain comes as a welcome surprise. Drinking something hot crosses our minds. Along the tundra bench I gather some fragrant blossoms and leaves from the labrador tea plant (also known as Hudson's Bay tea, or *Ledum palustre*). Dennis and I grew quite fond of labrador tea while living in Arctic Village. Local village residents introduced us to the tea, which is usually harvested in the fall when the aromatic blossoms are dry. We like to brew it in small quantities with black tea, having discovered that drinking too much straight labrador tea disrupted our systems and kept us on the run.

After tea, crackers and peanut butter, and plenty of gorp, we fall asleep to the soft pattering of our first rain. Tomorrow, after the restfulness of deep, rain sleep and intermittent tent reading, we plan to hike some seven miles to the Okpilak's ice-mantled headwaters.

The next morning we awaken to partly cloudy skies and cooler temperatures. We're anxious to explore the upper reaches of the valley and wonder if any other people have attempted our planned route. We knew of no one who had walked this particular route, and none of us had seen the upper region of the Okpilak from the air. The unknown makes it all the more exciting. With a light lunch, crampons, ropes, jumars, and carabiners, we soon take off with ice axes in hand.

Without the backpacks the pace seems fairly quick through a landscape that is impressively raw. For several miles we pick our way around unstable rocks along undulating talus slopes and across lateral moraines. The rock hopping is a killer on the feet. Occasionally we find comforting patches of dryas-specked tundra, which gives our feet the kind of relief one feels on roller skates when transitioning from rough asphalt to a smooth sidewalk.

The closer we get to the Okpilak's hidden glacier, the more rocky and rugged the terrain. We frequently cross ankle-deep streams that have incised the talus slopes before pouring into the Okpilak's gorge. Having had enough of the rock scrambling, I'm eager to rope up and enter that exquisite, naked world of snow and ice. It has been a year since Duncan, Dennis, and I climbed the glaciers and lofty peaks of the Canadian Selkirk Range. We're hungry for glacier climbing just as a kayaker grows anxious for white water after storing his boat in the dust of winter.

In a few hours we reach the ten- to fifteen-foot-high snout of the receding half-mile-wide glacier. Numerous rivulets of glacial meltwater cascade off the glacier's brow. Some of the rivulets flow into turquoise blue caverns where large chunks of ice have melted or calved into the riverbed.

Perhaps what is most striking is a tributary glacier that merges with the Okpilak. It is shorter and narrower, yet we are dwarfed by its enormous snout, perhaps fifty to seventy-five feet high. The impressive wall of ice reveals countless annual snowfall lines, like tree rings. We stare at the giant ice wall for some time, looking back through the years of heavy and light snowfalls. Weather records on ice.

We strap on our crampons, rope up, and climb onto the river of ice. All of the snow has melted on the lower end of the glacier, so our crampons jab into pure ice, making a loud crinkling sound, as though we are walking on shattered glass. The open, obstacle-free arm of ice is liberating. No longer are we picking our way through the talus or tussocks, and we don't have to skirt around crevasses yet. There is great freedom to view the surrounding glacier-gouged valley walls, breathe the pure mountain-borne air, and stride across an enormous ribbon of ice that perhaps no human has crossed.

The caribou, however, have been here. Not far up the glacier we pass two sets of caribou antlers, the first signs of animal life. Caribou have that habit of seeking out high, windswept ridges and mountains. When caribou are harassed by insects in the summer, the high country, like the coastline, provides cooler, breezier conditions and fewer bugs. During the winter months windblown ridges expose lichens, the caribou's main food source. The open high country also provides good visibility for spotting wolves.

Throughout the field of ice are numerous glacial rivulets, no wider than my hand, snaking through the ice. The tiny troughs of liquid possess that magical light-blue color seen only on glaciers. It is a hue so light and pure that you think this must be where blue is born. It is a color like the icy blue of some huskies' eyes, but more translucent. We pause to drink some of the ice water, which is so cold that it hurts our throats when swallowed.

Continuing up the glacier, we walk between two small medial moraines. Small pieces of rock and gravel pepper the ice. Each pebble has absorbed solar heat and melted out its own individual ice pocket. It is as though we are walking on a giant frozen sponge, with thousands of tiny black holes. Beneath the pebble-specked tongue of ice, through a small fracture in the glacier, we can hear the distant, hollow sound of flowing water.

Soon we reach the firm line of the glacier and begin to plod through several inches of wet snow. The more elevation we gain, the deeper the snow becomes. Dennis and Duncan take turns leading, each probing for crevasses with their ice axes. The deeper we sink into the snow, the more anxious they become.

"I don't like these conditions. I can't tell if I'm going to sink into the snow and bottom out, or fall into a hidden crevasse," Dennis says.

As we near the pass, Dennis and Duncan begin to sink up to their hips, each time not knowing whether there is a gaping crevasse beneath them. Alison and I ready our ice axes and prepare to self-arrest in the event one of them disappears beneath the snow. We had all practiced the self-arrest safety technique on other glaciers but had never had to execute the measure in a life-and-death situation. I feel ready for the maneuver, but nervous.

"There's nothing but air beneath here!" Duncan says in an alarmed voice. His ice ax is fully extended, well beneath the snow layer.

We backtrack and zigzag along a route that avoids the crevasse. The probing and trudging through the snow are time consuming. Dennis and Duncan are becoming exhausted from sinking in and pulling out.

"Another one!" Dennis shouts.

Within a mile of the pass, we had clearly entered an area full of crevasses. With the poor snow conditions and the dangerous element of invisible crevasses, Dennis and Duncan recommend that we turn back. It is not worth the risk of one or some of us plunging into crevasses in such a remote location. They believe it is almost impossible to probe out the remaining mile safely, and we still have seven miles to travel back to camp.

Although we are disappointed about turning around, we still celebrate an important aspect of the trip. For the first time we walked from the mouth of one of America's northernmost rivers to its source in the heart of the Brooks Range! For the last seventy miles we had drunk, waded through, bathed in, and slept by the waters of the Okpilak drainage. We had reached the birthplace of this glacial-fed river.

The next day we hobble around camp with a few new aches and pains, the most noticeable being sore feet. Talus slopes coupled with crampon walking had beaten our pads. We discuss our options for travel as we eat our granola and freeze-dried strawberries. Because of the unstable glacier conditions, we decide to eliminate the option of crossing the pass into the Hulahula drainage. When we consider our schedule for planned food drops by two different bush pilots, the wall of glacier-covered peaks between our location and the Hulahula, and the raging Okpilak, we see no feasible route to get to the Hulahula River without considerable backtracking. There is a chance that we can find a place for Walt to land on the Okpilak; then perhaps he can ferry us over to the Hulahula drainage. He is scheduled to give us an air drop in a couple of days, and perhaps we can flag him down.

Later in the morning we make the five-mile return hike to our food cache at Isto Valley. When we reach Isto's tributary in the afternoon, the dark gray waters are raging and the river boulders are rumbling once again. We are out of food and getting hungry. Anxious to cross the river and set up camp, Duncan decides to be the guinea pig and go for it. He barely makes it across, staggering in several places, with the water hitting as high as his waist at times.

Dennis, Alison, and I decide to use the rope and look for a better spot. Huge troughs had formed in the tributary, and the thunderous sound of rolling rocks worries me. If one of us slips in that water, it will be nearly impossible to regain proper footing, especially when weighted down by a pack. The water is near freezing, and there is always the chance of being swept into the main channel where one's head might hit a rock, and a body numbed by the cold water could easily drown. I don't like this one bit.

I belay Dennis as he crosses one section of the river. He seems surefooted during the first third of the river crossing. Then he loses his footing and begins to sway back and forth, trying to regain his balance. The more he sways, the greater his chance of falling. Suddenly he is in the water, with his submerged pack pulling him down into the rush of water. For a moment I can see only his arms waving and hands grasping onto the rope. The force of the current has me pulling the rope with all my strength, the kind of strength that you don't know you possess until you are flooded with adrenaline.

Within a few long minutes Dennis is pulled safely back toward shore, his head above water and his feet pushing off river boulders through the surging water. He lunges out onto the riverbank looking half frozen. Shivering and soaking wet, Dennis suggests that he try to cross a wider, shallower section of the river, one closer to the Okpilak's main channel. I suggest that he dry out and wait until morning, until the river drops. Dennis argues that the food is on the other side of the river and that he doesn't want to get wet twice. I don't like it, but I give in.

He heads out a second time; I anxiously hold the rope as taut as possible. One step . . . two steps . . . three. Soon he's in the middle of the channel. He cautiously moves through the swiftest, most dangerous section. He sways for a few seconds, then continues on to the other shore.

"This place is much easier to cross. You two can do it," Dennis yells to Alison and me. We look at each other and aren't quite so sure.

With Dennis belaying, it's my turn. I step into the icy water and feel it soak through my wool socks down to my toes. After a few steps I feel the tremendous rush of water. My legs quickly grow numb. I know I can't hurry, but I mustn't waste time. If my legs and feet become too numb, I'll have a harder time feeling my way. Loose boulders continually roll by me. Never have I crossed such a cold, fast-moving river.

The river is raging so swiftly that I can't look down, only forward. The few times when I do look at the water, I get dizzy and start to lose my sense of balance. In the middle of the channel I feel as if the river is about to swallow me. I've lost all feeling in my legs and can't seem to move them. My feet are solidly planted, and I can't raise my numb leg through the current. I have the horrible feeling that if I do lift my leg, the river will take it away.

Something has to change. I try angling downstream with the current instead of fighting my way across it. It works. Soon I reach the opposite shore. My legs and feet begin to throb as my blood circulates. In a few more minutes Alison safely crosses the river along the same path.

The following day Dennis finds a long, level spot on the bank of the Okpilak where Walt can land his Cub. We clear the strip of all large rocks and flag the ends of the area with red bandannas. We plan to shoot up a flare when we see the Cub, signaling Walt for a pickup.

During the remainder of the day we relax around camp, reflecting on the first two-week portion of our trip. The fact that we were unable to scale Mount Michelson or cross the pass into the Hulahula drainage brought disappointment, but at the same time we felt great adventure in our efforts to meet the challenges on this truly wild land. When I think of other glaciated peaks we have climbed, outside Alaska, there was always adventure, but fewer unknowns. The approaches to mountains such as Glacier Peak, Mount Baker, and Mount Rainier in the Cascades were along designated trail systems. Instead of our fording swift rivers, there were often bridge crossings. It was common to find glacier trails of other climbing parties leading to the summit. And if needed, you could always consult with some mountaineer about your planned route.

As it turned out, Dennis and I would reach the summit of Mount Michelson via the Hulahula River two years later. From Michelson's summit we would look down upon the Okpilak River and reminisce about this trip. In one far-reaching glance, we would see the sweeping coastal plain touch the Beaufort Sea, and the great stretch of pack ice beyond, curving toward the North Pole. We would watch Dall sheep graze upon velvet green foothills below us, see silver rivers wind through glacier-carved valleys, and view countless snow-mantled peaks cutting their summits into the bluest of skies.

We would remember these days on the Okpilak River and our anxiousness to climb Mount Michelson. More important, our long-awaited ascent of Michelson would become more meaningful because we had worked hard, twice, to get there.

7

The Hulahula and Chandalar Rivers

WALT GIVES THE CUB FULL THROTTLE, gunning it down the bumpy stretch of tundra in what seems a fury. I sit nervously in the back seat as the rattling plane jolts its way toward the river. I lean as far forward as possible, hoping that a little extra weight in front will help the plane lift off. Then again, if we crash, maybe the blow won't be so hard if my head is buried between my knees. Walt's words echo in my mind. When he first landed on our hand-groomed strip, he told us that the landing site was marginal, less than three hundred feet long. But he *thought* he could ferry us to the Hulahula River, one by one, with a little help from the wind.

Today there is a north breeze, and apparently it is enough. I open my eyes when I feel that wonderful sensation of being airborne. Looking out the window, I see the Okpilak River boil beneath the Cub at what looks like a safe distance. Walt gradually banks the plane and steadily gains altitude. My heart stops pounding.

Before long he turns up the Okpilak's west fork, and we gain enough altitude to see some of the glacial-mantled mountains surrounding Tugak Peak. The lower end of the Okpilak's west-fork glacier comes into view, and its massive arm of ice curves around the same rib of mountains we had passed to the east. Several smaller tributary glaciers spill into the main glacier, like branches into their trunk. We are surrounded by cliff-hanging rivers of ice that curve around mountains, and fall toward rock-jammed valleys and canyons. The spectacular views are all going by too fast, even in the slow-moving Cub.

The higher we climb, the more bumpy it gets. Walt must clear a 7,000- to 8,000-foot ridge to drop over into the Hulahula drainage. The closer we get to the ridge, the stronger the downdrafts and the more bouncy. The Cub's altimeter is jumping all over the place, and a couple of times my head bangs into the fuselage. It doesn't look like we're going to gain enough altitude to clear the rock wall in front of us.

Walt keeps the nose up and gains a few hundred feet more elevation. Every so often the bottom falls out, and we're swallowed by another downdraft. My stomach jumps to my throat. Finally, the roller coaster flight ends. Suddenly we are there, just above the ridge line. If the window were open, I could reach down and almost touch the serrated ridge.

It's smooth sailing now as the Cub glides like an eagle, dropping down along East Patuk Creek and beyond, to the broad emerald-green Hulahula River valley. What a dramatic change of landscape! Unlike the Okpilak's confining drainage, with its steep valley walls, talus slopes, and gorges, the Hulahula region is full of open valleys, sweeping tundra slopes, and relatively unobstructed views. Where the Okpilak River valley is less than one-half mile wide in some sections, the Hulahula River valley is closer to two miles in breadth.

Walt lands on a gravel bar on the Hulahula, just south of East Patuk Creek. Dennis is there waiting, and we soon set up camp near the gentle river. I had forgotten the soothing sound of a slow-moving river after two weeks of camping on the turbulent Okpilak. After setting up the tent on a flat bench of tundra, we explore the gravel bars. The Hulahula is so clear that the river stones in shallow areas appear as if they are seen through a magnifying glass. No longer will we have to wait hours for the silt to settle out of our drinking water.

Judging by the variety of tracks we discover, the gravel bars must be teeming with life. Wolf, ground squirrel, and weasel tracks crisscross the bars. We find an old set of moose tracks and piles of droppings near a stand of willow bushes. By the river's edge are several sets of ewe and lamb tracks. These Dall sheep must have dropped down from the high country for a cool drink.

Within an hour we spot numerous sheep grazing on the upper valley slopes that surround us. On one ridge we count twenty-six rams, with varying curl lengths. Some graze, others rest on grassy benches, and a few stately animals walk along the ridge crest. Their horns curl into a sky the color of arctic forget-me-nots.

The Arctic Refuge provides excellent summer and winter habitat for Dall sheep. Of the estimated 30,000 sheep in the Brooks Range, approximately 11,000 live within this refuge. The vast majority of these sheep are found on the north side of the Brooks Range along high-density drainages such as the Hulahula, Canning, and Kongakut rivers.

Biologists believe that Dall sheep prefer living on the north side of both the Brooks Range and the Alaska Range because those sheltered regions receive less annual snowfall from winter storms. By living in the snow shadow of these mountains, the sheep have greater mobility and more readily available forage. North–south drainages, such as the Hulahula River, are particularly attractive to sheep because of strong winds that funnel down the valleys through low-lying passes and scour the valley ridges. The windblown ridges expose desirable vegetation, such as sage in the East Patuk drainage, for the sheep's winter diet.

The following day we decide to explore a 6,000-foot ridge on the west side of the river. It's a perfect summer day, with continued warm temperatures and puffy cumulous clouds scattered above the mountaintops. The more we gain in elevation, the more spectacular the vista. All of us stop frequently, not because of being winded, but to gaze down the valley and see what new mountains have popped into view. We are now about ten miles from the Continental Divide and can see the jagged peaks and fluted ridges that form that east–west rampart. Yet the pastoral setting of the broad Hulahula valley softens the rugged high country. The numerous tributary streams seem to caress the mountains rather than carve through them.

Like the Okpilak River valley, the Hulahula drainage was also swallowed by glaciers. Here, however, the glaciers apparently spent more time scouting and scooping out the valley. Near the northern foothills along the Hulahula River you can see the remains of two enormous terminal moraines. Where the narrow Okpilak valley is dogged with lateral moraines and talus, the Hulahula's more mature glacial system has pushed most of its debris out to the southern edge of the coastal plain.

Farther up the ridge Duncan spots the white coats of several Dall sheep rams against some dark outcroppings. Eight to ten sheep run along narrow ledges and climb up cliffs with the surefootedness and muscular legs of a sprint runner. Soon they disappear behind the small rocky fortress. A few moments later, one lone scout appears at the top of a cliff and studies us. The young-looking ram watches us for a good half hour, until we vanish over the next ridge.

Sheep-hunting season will soon open in the Arctic Refuge, and these sheep certainly seem wary. Each year many ram hunters come to the refuge. The hunters are often accompanied by guides from one of several commercial operations. As much as 70 percent of the annual Brooks Range sheep harvest comes from within the Arctic Refuge. The hunting of sheep and other species is so popular that many seasonal visitors to the refuge are hunters.

Soon we stroll along the crest of a ridge that offers one of the most remarkable views. Beyond the Hulahula valley, Mount Michelson, Mount Chamberlin, and the many nameless glaciated peaks around the headwaters of the Okpilak lift their snow-masked faces skyward. Below, there are countless green valleys sheltered beneath towering ridges and peaks. Each valley possesses its glistening ribbon of water that drains into the Hulahula. Flower-specked tundra slopes roll gently toward the streams and seem to melt into one another. The 360-degree panoramic view is so spectacular that it is impossible for the eye to contain it all. I turn in a slow, unending circle for several minutes to soak in the beauty in the pure, unbroken silence.

Farther on we stop at a beautiful narrow lake that is slivered between two ridges. The silver-blue lake is perhaps the remnant doorstep of a glacier that once funneled through this valley. We walk around an outlet that trickles out of the north side of the lake. The tiny stream is laced with alpine grasses that are peaking in the green of summer. Sprinkled in the grasses we find the soft lavender of Jacob's ladder mixed with pink bistort, fireweed, and yellow *Oxytropis*. The setting here is reminiscent of some of the alpine lake country we've visited in the high Sierra: a land of gentle streams, open green meadows, and mirror lakes.

Heading back to camp, we come across many wolf tracks and old signs of grizzly diggings. I walk past one large bed of Alaska *Boykinia*, commonly known as bear flower, where a grizzly has browsed many of the plants. Bear flower is one of the taller members of the saxifrage family, growing up to three feet in height, with numerous creamy white flowers growing on its long branching stalk. The flowers, stalk, and leaves of the plant are a favorite bear food.

Near the river again, I pass a large pingo full of ground squirrel holes. One side of the pingo has been tilled up by a grizzly who was obviously in hot pursuit of a mouth-watering squirrel. How frightful it must be for a squirrel to be madly running through its tunnel system to escape the gouging claws of a grizzly. As I walk on, a few gregarious squirrels pop out of their holes, making their chee . . . chee sounds at me.

The next morning we backpack south along the braided Hulahula River in sunbeating temperatures. It is early August, yet we continue to sweat our way toward the divide. There is one element of relief. We've noticed few mosquitoes in the Hulahula drainage, and that means they must be nearing the end of their life cycle. There is great joy in not having to smear on that skin-eating jungle juice.

Roasted by the sun, I frequently stop along the river to splash water on my face, drink the sweet water, and drench the bandanna worn around my neck. There is nothing like an ice-cold bandanna on the nape of the neck on a day such as this. It seems to lower the body temperature by sending a momentary chill down the spine. The only trouble with these cooling stops is the starting up again. Swinging on that lead-heavy pack while maintaining my balance is a chore. Sometimes my body just collapses, like a noodle, on the tundra. The key to such falls is to not get frustrated and curse at my pack, but, instead, to pull out my binoculars, lie flat on my backpack, and glass for Dall sheep.

After we group together for lunch, Duncan takes the lead and strides out as fresh as ever. I move as though in slow motion, lifting each boot just enough not to trip. Duncan is about one hundred yards ahead of me when he suddenly stops dead in his tracks near a small gully with spotty willow bushes. Looking startled, Duncan backpedals, then trips over a willow bush, landing flat on his back. As he falls, the rest of us see a pale wolf emerge from the gully.

A wolf! All of us are stunned at this first sighting of a wolf in the wild. None of us utters a word as we watch the beautiful animal circle helpless-looking Duncan, only a few yards from his prone body. With its tail up, the wolf walks around Duncan and stares at him with penetrating eyes. Duncan watches the wolf but doesn't budge. The animal seems to be panting hard in the heat of the day and appears lethargic. After a few moments the wolf trots up the valley a short distance then turns around to take another look at Duncan. Farther on, the wolf joins a dark gray wolf, and they continue up the valley. Once the wolves have moved a good distance away, we all join Duncan, and he explains what had happened.

"I saw this thing jumping out of the gully in front of me. I expected it to be a bear. When I fell backward, the backpack pinned me to the ground. I felt totally helpless. Then I saw it was a wolf and thought to myself, what now?"

Duncan said that once he had established eye contact with the wolf he felt less scared. The wolf seemed clearly in control of the situation, and Duncan just lay there in a nonthreatening posture. He sensed that the wolf was only watching him and would not attack him. Duncan had apparently startled the wolf from a midafternoon nap in the shaded gully.

Later in the afternoon we set up camp, near Itkillik Creek, which means "Indian Creek" in Inupiaq. Dennis and I are so hot we soak ourselves, clothes and all, in the cooling, shin deep river. The undulating tundra slopes near camp are covered with low-bush cranberries, so we spend the rest of the afternoon stocking up on berries for our bannock and pancakes.

In the evening we watch lengthening mountain shadows drape the surrounding valleys. The curtains of dusk slowly inch their way toward our camp and soon there is only sunshine on the distant peaks. As the light grows more dim, we crawl into our tents, exhausted from the long, hot day. To our great surprise we receive one more gift of wilderness as we zip our sleeping bags closed. A new musical group serenades us to sleep. Up the east fork of the Hulahula River, we hear the songs of several wolves ringing the calm air. I had always read in books that wolves howl, but tonight they sound like they are singing. Each wolf has its own range of wailing, yet melodious, cries. I drift off to sleep feeling privileged to sleep in the backyard of this wolf pack.

Morning dawns clear and cooler. We pack up our gear and look forward to setting out for the Continental Divide. Today we plan to hike about ten miles to the headwaters of the Hulahula River, over a gentle low-lying pass, and cross over to the headwaters of the East Fork of the Chandalar River.

The walk up to the pass is an easy stroll on smooth alpine tundra, following the jogs of this delicate-flowing portion of the Hulahula River. We gradually gain two thousand feet in elevation, which is hardly noticeable. Then again, perhaps we are all in better shape now. The closer we come to the pass, the narrower the valley, the more quiet-flowing the stream, and the steeper the limestone ridges that rise above us. As we did in the Okpilak River valley, we see many sheep trails etched across talus slopes. These rocky slopes are different; they are more broken up by grassy chutes that finger dawn through the talus, likely providing good sheep habitat.

Near the final bend in the stream, with the pass in sight, we spot a graceful golden eagle soaring above us. Its huge wingspan takes command of the entire sapphire-blue ceiling above us. An adult golden eagle averages about eleven pounds, with a wingspan of six to seven feet. Golden eagles have tremendous strength harnessed in their feet and deadly talons.

These eagles prey largely on small animals such as the ground squirrel and ptarmigan, but they are also known to successfully attack and kill Dall lambs and caribou calves by diving upon the back of the animal and piercing vital organs. In some years golden eagles, particularly nonbreeding birds under five years of age, have been documented to kill more radio-collared caribou calves within the Porcupine caribou herd than wolves or grizzlies. Calves killed by golden eagles within the refuge have weighed as much as forty pounds.

Just before reaching the pass, Dennis and I pause to listen to the soft trickle that marks the birthplace of this northern river. The water's sound is as delicate as the fluttering sound of quaking aspen leaves. In our noisy world such subtle sounds in nature are savored, particularly when they are heard in a quiet area of the planet. In our backyard I may hear the rustling sound of aspen leaves, or a brook running through our property, but it is not the same. I will most likely hear traffic on the road, airplanes overhead, motorboats on the river, or the ubiquitous lawn mower. Here such gentle sounds stand out like the clean edge of a full moon against a black winter sky.

Another two hundred feet of elevation and we are standing on Guilbeau Pass, named after a geologist who attempted to walk north from Arctic Village over the Brooks Range via the East Fork of the Chandalar and Hulahula rivers. Guilbeau set out alone one summer in the early 1970s and planned to study the geology of the area. He apparently ran out of food and mysteriously died on a north-facing glacier below the summit of Mount Michelson.

From the crest of the divide, we have an impressive view. Looking west, we can see the origin of the Hulahula River and follow its winding course down the valley until it hairpins to the north and disappears. Looking east, we gaze down upon a deep blue lake, a rippling jewel cradled between dusty gray mountains and ridges that form the V-shaped valley. The lake is yet another birthplace, this one for the south-flowing East Fork of the Chandalar.

Until now we have camped by north-flowing waters that empty into the Beaufort Sea, and then eventually into the Arctic Ocean. Now that will all change. We will follow the Chandalar River, knowing that its waters spill into the Yukon River, pass many Native settlements, and eventually flow into the Bering Sea.

We set up camp near the edge of the lake and soon discover a group of sheep just above our camp. While watching them in the evening, a southeast wind picks up, and we all sense a change in the weather. As the sun falls behind the mountains, the dew point drops quickly, and a thick mist begins to magically envelop us. It is as though we are witnessing the birth of a cloud from the inside looking out.

For one day and two evenings we are swallowed by heavy mist and fog in our tents. Although our nylon cocoons keep us somewhat warm, the fog and intermittent light rain soak our bags, clothes, and gear. Most of the time we can see only a trace of the lake that is within a few feet of our tent door. The view gives us the feeling of being in a storm at sea. Since the mood is right, I read *Moby Dick*.

On the second morning a sliver of sun creeps into our tent. Outside, the drenching white is evaporating as quickly as it had saturated us. Within a few minutes we can see the surrounding mountains through the glitter of a fine mist. By the time we finish breakfast the sun is boldly shining upon us again; yet, unlike past days, there's a hint of autumn in the air.

Descending from the pass, we soon reach a section of the creek that cuts through a narrow gorge. Dennis and I decide to walk, or wade, our way down the canyon, while Duncan and Alison hike above us along the ridge. The creek winds through unusual formations of rippled and twisted sedimentary rocks. On the creekbed, we find many coral fossils embedded in gray rocks. The Lisburne corals of the eastern Brooks Range are about 300 to 350 million years old. These corals are reportedly a different and younger species from the coral found in the central Brooks Range. The age of these fossils suggests that the eastern portion of the Brooks Range was covered by seas at a later date than the central Brooks Range. Again, it's hard to imagine that we are walking in a zone that once was a marine environment.

Soon the canyon opens up, and we can see the Chandalar's broad river valley below us. We drop down to the main river valley to a point where several tributaries merge with the main river. Dennis spots a pie-shaped tundra slope that is wedged

Chandalar River valley

between two major forks of the Chandalar. The slope offers a perfect campsite and a great view looking down the middle of the Chandalar's U-shaped valley. And, at last, there are real, live trees to look at as well!

Nearing our destination, we approach giant-looking stands of willows. When compared to the knee-high willows on the north side of the divide, these ten- to fifteen-foot-high shrubby trees seem to tower over us. After looking down at plants for so many miles, it's pleasant to look skyward through rustling leaves. Many of the willows have had some of their leaves stripped off by browsing moose. Some trees are hedged, their branches uniformly nipped off at the same height. It looks as though the moose circled around the tree with clipping shears.

In the main valley we see caribou and moose tracks all over the gravel bars and along the banks of the river. Looking south, we spot a huge sheet of *aufeis*, or overflow ice, that bridges most of the river. During the winter *aufeis* forms as water runs over a frozen section of the river and freezes. The overflow water is often generated by underground springs, and layer up on layer of overflow ice can build up on the river through the course of a winter.

Some *aufeis* grows to be numerous feet thick and stretches for miles in length and breadth. Depending on temperature and location, much, or all, of the *aufeis* can melt during the summer. Also, there are areas where thick sheets of ice can remain for many years. I like to think of perennial *aufeis* as river glaciers.

From camp we hear an occasional rumble as *aufeis* sections calve into the river. The booms sound like distant claps of thunder. Dennis and I decide to hike down to the ice for a closer look. Just as we reach the mile-wide section of river ice, we watch a huge chunk of overhanging ice calve into the river. The wedge of ice makes a terrific splash, and a wake of backwater flows upstream for several hundred feet. The ice chunk had broken off where the river's main channel cuts through the blue-tinted sheet of ice. The surface of the *aufeis* is pocked with tiny faint blue pools surrounded by rivulets of meltwater that cascade off the ice sheet. The perimeter of ice is impressive, measuring twelve feet thick in some places.

During the next few days the weather dramatically changes. Instead of being baked out of our tents, we begin our mornings in the dampness of fall mist. Ominous gray clouds hang over the river valley, and a chilling wind makes me pull the wool out of my pack for the first time. Plants, like the prolific cotton grass, disperse their next generation of wind loving seeds across a tundra flooded with autumn color. The bearberry is turning a deep crimson, the blueberry bushes pale red and orange, the willow trees yellow and gold. It is New England's autumn in miniature. But it is not October. Autumn arrives in the Far North in mid-August.

A few days later we continue our southern journey beneath gray and white billowy clouds, with the sun occasionally beaming heaven's light through them. In the distance we can see the beginning of the dark spruce forest stretching across the valley's floor. The forest is cradled by 5,000- to 6,000-foot treeless mountains and ridges whose naked profiles tower above the valley of green. The striking lines of the pale to slate gray mountains seem to enhance the forest. The forest, in turn, warms the bases of the mountain ridges and accentuates the vertical relief.

Soon we reach the first small patch of white spruce trees, which are up to three feet in height. There appears to be one maternal tree with a gnarly stump three inches in diameter. Surrounding this most northern pioneer are five to six hardy seedlings. These dwarfed trees, along the northern limit of spruce in the Brook's Range are the oldest white spruce trees within interior Alaska. They take many decades to grow because of the short growing season and the stressful, harsh climate. Because of the cold climate, however, these twisted trees are not subject to the kinds of diseases or decay encountered by trees of more temperate, bug infested zones. A small tree, such as this one, may be several hundred years old.

After hiking another mile or two, we come to a golden meadow that is covered with the bobbing, wispy heads of cotton grass. Just beyond the meadow, along the main channel of the river, we see the first large stand of spruce trees, flanked by formidable mountains that rise several thousand feet above us. I decide to stop and absorb this dynamic transition zone. I start to pull out my camera, but then realize a photograph will not capture the immensity of the valley or the emotion of this scene. Instead, I unearth my watercolor set and try to paint the feeling of this place. It's easy to snap a picture, but I remember a place more vividly if I take the time to study the landscape and sketch the scene.

After finishing the painting, I walk on toward the edge of the forest. Just before I head into the woods, something dark brown moves in the willows in front of me. Drawing closer, I can see the quills of a porcupine rustling against the willow leaves. This animal seems unalarmed by my presence as I stand within twenty to thirty feet of its fortress of needles. It is busy nibbling on willow leaves. Occasionally two curious brown eyes glance up at me, then the thorny animal waddles a few feet to another willow bush. Its small, squarish, black head and tiny mouth look like a cross between a raccoon and a coatimundi.

After a time the porcupine slowly moves to a nearby creek, its quills rippling along as if blown by the wind. I keep a friendly distance. It dips its black head into the water, and there's a faint sucking sound as it breathes in a drink. I kneel down for a drink as well, then wish the porcupine a good day and head on to join the others.

With just a few strides I enter an entirely different world. The smell of evergreens! My feet spring off the spongy moss-covered, needle-laced floor of the spruce forest, and a multitude of singing birds greets me. Boreal chickadees whistle their cheery songs, gray jays melodiously chatter to one another, and ravens caw overhead. A northern shrike silently rests on the crown of a tree, looking out over the meadow. The trees here are in the ten- to twenty-five-foot range, but they appear to be giants in comparison to the dwarfed vegetation on the north side of the Brooks Range.

Near the edge of the forest I spot Dennis, Duncan, and Alison looking at something on the ground. They've discovered what looks like an old cache that has sunken into the mossy tundra. The rectangular cache, about five feet long and two feet wide, is built of spruce logs and sits a few feet above the ground. The cache was used perhaps by an Athabaskan hunter or trader. The Chandalar–Hulahula River system was one of several historic trading routes used to reach Inupiat settlements near Barter Island along the Beaufort Sea coast.

The coastal zone near Barter Island was one of the main centers for trading activity. Through an extensive trading network that stretched across Alaska and into Canada, Inupiat traders would travel to the Barter Island vicinity from distant places such as the McKenzie River delta in Canada, and the Colville River region to the west. Athabaskan traders would travel north to the coast from the Arctic Village region, and Inupiat traders would also visit the Arctic Village area for trade meetings.

Visiting Natives commonly had trading partners with whom they exchanged goods such as furs, pots, knives, tobacco, and other items. In the late nineteenth century, whalers introduced rifles and ammunition, which became desirable trading goods for Athabaskans. According to several sources, the first cartridge rifles used by Arctic Village ancestors came via the Inupiat on the arctic coast.

Maggie Gilbert, of Arctic Village, now deceased, once told me that the two trading items most desired by the Inupiat were wolverine skins, which are highly water resistant, and white spruce pitch, for medicinal purposes. The Athabaskan Gwich'in, of northeastern Alaska, still apply spruce pitch to wounds and infections, and use the pitch mixed with water as a cough medicine and as a remedy for other illnesses. Moses Sam, of Arctic Village, once applied spruce pitch to an infected knife cut on Dennis's hand, and the swollen infection healed quickly. Since spruce trees are not readily available in Inupiat territory, Athabaskan traders traditionally carried a ball of spruce pitch wrapped in a cloth when they traveled north to trade.

Nomadic ancestors of Arctic Village residents would travel to the arctic coast by dog team in the winter and on foot in the summer. According to written accounts of the 1860s, the 300-mile round-trip by dog team took about ten days. During the summer months, until about the 1930s, elders in Arctic Village remember traveling with their parents and extended family into the Brooks Range to spend several weeks hunting for sheep and other game. Dogs carried supplies in hand-sewn caribou-skin packs.

While mothers, children, and some members of the family stayed at these hunting camps, other members of the family would walk on to the coast to trade. Isaac Tritt, an Arctic Village elder, remembers walking as a boy with his father to Gordon's trading post at Demarcation Bay, near the Canadian border. They traveled another northern route following the Shiinjik and Kongakut River drainages to the Beaufort Sea. (Shiinjik is the correct Gwich'in spelling, meaning "salmon river," although it has commonly been spelled Sheenjek for many years). The fact that such hardy travelers followed these trading routes for decades makes our recreational trek from the coast seem insignificant.

Later we set up camp near the edge of the woods overlooking the Chandalar. Dennis meanders down the braided river and catches a few arctic grayling while the rest of us pick wonderfully ripe blueberries, our tongues and teeth turning purple as we sample them. While wandering from bush to bush, Duncan wanders off and stumbles upon two dead ground squirrels that have been killed recently by a grizzly. One ground squirrel has been severed in half by grizzly teeth, as though a butcher's knife had sliced through it. All that remains is the bottom half of the squirrel. The other dead squirrel is still warm, with cheeks full of willow seeds. We see fresh bear tracks and scat near the kill site.

"I hope we didn't disturb his lunch," Duncan says in a worried voice.

We speculate that the bear smelled us and was scared off in the middle of its meal. Since the dead squirrels are close to camp, Duncan picks them up and carries them down to a gravel bar near the edge of the river. He figures that we might as well not provoke the bear by sleeping too close to the grub. However, I wonder if the bear will come back and be ticked off that the squirrels are gone, then seek revenge by ripping open our packs.

The next day we awaken to clear blue skies and no sign of the grizzly. All of us agree that it's a perfect day to climb a mountain and get an eagle's view of the valley and surrounding country. Our eyes focus on the summit of a mountain that's about twenty-five hundred feet above the river. We're anxious to gain some elevation, so we take the most direct route, ascending a series of steep gullies that are cluttered with loose scree. We stop often, out of breath, and gaze below at the spectacular valley. During one breather, Duncan spots a large, bull moose in the middle of the valley. His massive antlers and enormous body are easy to follow through the meadows and willow thickets.

Finally we get above the last of the scree and find ourselves on firm alpine tundra that is still in its peak of summer greenery. The familiar white heather laces the tundra, along with bright patches of cinquefoil and moss campion. We have momentarily left autumn on the valley's floor and now savor one last patch of summer in the Brooks Range.

When we reach the summit, the panoramic view is overwhelming. It is as though we are on board a hot-air balloon and can see in every direction without obstruction. To the north, our eyes follow the broad Chandalar valley to where it curves out of sight and continues to its origin. We can see the jagged Continental Divide and several remnant glaciers clinging to the surrounding mountains. Between this vantage point and the divide lie countless naked ridges, and so many nameless tributary creeks and valleys that one could spend a lifetime of summers and not see them all.

Below us, the valley is cloaked in the dark green of spruce, with the river shimmering between the edges of the forest and the golden willows. Looking south, we see the Chandalar valley widen even more, to perhaps four or five miles in breadth. The river becomes more braided, the spruce forest more dense; and chains of small lakes and teardrop-size ponds lie in the distance. It is a land that offers habitat for a new array of species, such as the muskrat, beaver, and marten. Some sixty miles to the south, we can see Datchanlee, the long ridge that exposes its top half above the spruce forests, just beyond Arctic Village. While hunting on that distant ridge with village residents, we saw our first view of this spectacular valley.

Along Datchanlee, which means "timberline" in Gwich'in, I can picture the canvas wall tents of many Arctic Village families, dotting the ridge near the edge of timber, with hunters nearby in search of the returning caribou. I can smell smoke from the campfire, the steaming black tea brewed in an old blackened coffee can, a caribou head roasting beneath a tripod of spruce branches, and the pungent scent of a ground squirrel's fur coat singed off by the fire.

I can picture young children with their fathers and mothers walking near camp, checking their ground squirrel traplines, and hunting for ducks at nearby ponds and lakes; old women, their weathered skin coffee-dark at summer's end, sitting in tents on caribou hides cushioned by underlying spruce boughs, sewing intricate beadwork designs on hand tanned whitened caribou hides, making caribou-skin boots and mittens for their grandchildren, and telling stories in Gwich'in that have been passed down from generation to generation.

I remember the stories of several elders who vividly recalled accounts of starvation during their lifetimes. There have been years when the caribou, the life and blood of these people for thousands of years, were not to be found. Elders remember the extremely cold winters where ptarmigan and fall-caught ground squirrels were the only sources of food for survival. It was not unusual to hear stories of some family members dying of starvation. There was a time, not so far in the past, when caribou hooves were used to make soup when there was nothing, or little, to eat. One day I saw an elder hanging hooves on a spruce tree near her camp, and I asked her what she planned to do with them. She told me that she still saved the hooves for survival, for "hard times."

It's not surprising that today's Arctic Village residents, along with Athabaskan residents in several other villages in Alaska and Canada, are vehemently opposed to the oil companies industrializing the calving grounds of the Porcupine caribou herd, a herd that has sustained their people and ancestors for at least ten to fifteen thousand years. The idea of tampering with a zone where tens of thousands of caribou bear their young is unthinkable to those who have depended on the herd; and there is scientific evidence that a major oil-field development, with the activities of up to six thousand people and the proposed web of roads, pipelines, and facilities, would indeed disturb the caribou. As predicted in the Department of Interior's 1002 report to Congress, the caribou would suffer the effects of habitat modification, reduced access to insect-relief habitat, and displacement from preferred calving areas. It is expected that such loss of habitat and related disturbances will change the distribution of the Porcupine caribou herd, and, the report notes, there is risk that a population decline could occur.

In the words of one resident, as spoken at a 1988 USFWS hearing in Arctic Village: "I don't depend on that damn oil. . . . I'd rather have my land right in my hand. I'd rather have piece of meat right on my table. That's what I've been raised up on. Not only me, but all these 100 people right in Arctic Village. . . . How are we going to live without caribou? . . . "

Another resident, Timothy Sam, spoke of conserving resources for the future: "Let's save it. You got kids. . . . You got to think about them. If we use up everything: our resources, the oil, and so on . . . what will our kids use in the future? Let's not deal with now. Let's deal with tomorrow or the next day. We don't have to go up there and drill it. We got no business. We got kids . . . ."

With binoculars I now scan the floor of the Chandalar valley and spot the remnants of an old caribou fence, or corral, built by the ancestors of Arctic Village residents. Caribou fences, commonly built in the shape of a keyhole, were commonly used until the turn of the century to capture caribou. During the 1890s rifles became commonly used, and the need for funnelling migratory caribou into an enclosure and catching them with the use of set snares gradually diminished.

Caribou fences took many days to construct, and several families often worked together to build them. The fence railings were built of spruce poles, which were linked together by supporting tripods. Women and children would collect willow or spruce roots, which were then soaked, stripped of bark, and boiled in coiled bundles. The pliable roots were useful for lashing together the tripods.

Many of the remnants of these fences are still visible from the ground and air. Between the years 1971 and 1973, forty-six fences were located in northeastern Alaska and the northern Yukon Territory under an Arctic Gas study prepared by Renewable Resources Consulting Services. Most of these fences averaged one to several miles in length, as measured from the open-ended base of the enclosure to the head of the keyhole. The fences were designed to intercept caribou along common migration routes.

Each fall, families would gather to drive the caribou into the corrals. As groups of caribou were spotted, the men would spook them into the corral by making noises and waving their arms. Once the animals were inside the enclosure, the entrance was blocked off by men, women, and children, and cut brush was used to close off the area. Sometimes the frightened animals would break through the fence or stampede back through the entrance.

Inside the corral the caribou became caught in a series of snares that had been constructed of braided strips of caribou hide and set on support poles. As a caribou's antlers or head became caught in the snare, hunters would move in quickly to kill the animal with a spear or bow and arrow. The animals were then removed from the corral to be butchered. Later on, families took on the task of repairing any breaks in the fence and replacing damaged support poles.

If families were successful in securing enough meat for their winter supply, they might camp at the fence site throughout the winter. Although fish, fowl, moose, sheep, and smaller mammals, such as the ground squirrel, have traditionally been important food sources for the Athabaskan Gwich'in, caribou has clearly been the mainstay, and still is today. In addition to being the principal food, the caribou has many other significant uses: skins are used for ground cloths and bedding; boots, mittens, and other clothing are made from tanned skins; tools are crafted from the legbones; rawhide rope, known as *babiche*, is made from braided strips of tanned hides; sinew line is made from the sinew that runs along the back of the caribou; and a caribou's toenail can provide the perfect protective base for a knife or scissors case.

While living in Arctic Village, I was deeply impressed by the extent to which our Athabaskan neighbors utilized the caribou. Every part of the animal had some sort of dietary or practical use. Hunting caribou meant far more than just meat on the table. The animal has represented a way of life for this rugged group of people for thousands of years.

The infinite view of valleys, mountains, and rivers, in such intense silence, would linger in my mind for days. Tomorrow we would reach Red Sheep Creek, marking the end of our long trek. There we would be surprised by a visit from the Allen Tritt family of Arctic Village. We had grown quite fond of the Tritts, having taught four of their daughters, and having learned much about the Athabaskan culture through Allen and his wife, Margaret. We would camp with them for a few days at Red Sheep Creek before returning by plane to Arctic Village to begin our new teaching year.

Dennis and I leisurely walk down to the valley and reflect on the many miles we had walked, the wide-ranging terrains we had experienced, and the diversity of wildlife we had encountered. Through the course of our month-long trip we had seen no other hikers. With the exception of occasional airplanes, our only discovered trace of humans was the old boiler near the Okpilak's fork, and the sunken log cache and remnant caribou fences in the Chandalar valley. There were no footprints, no sings of old campsites, no candy bar wrappers, no plastic. Even if others had walked our exact route, there was still that exhilarating sensation that we may have walked in places where perhaps no human had ever set foot. It is this spirit of pure wildness, in such an awesome land, that lingers on in our hearts and minds.

Part Three

ALONG THE DIVIDE

8

A Mystery Solved

A FEW WINTERS AFTER OUR TRIP across the Brooks Range, I'm pouring through a stack of topographic maps inside a twelve-by-fifteen-foot log cabin southwest of Artic Village. With so little daylight and forty-below-zero temperatures, it's a perfect time to dream of summer backpacking trips. Ever since crossing the Brooks Range via the Okpilak, Hulahula, and Chandalar drainages, I've longed to hike a route through the refuge that would follow different northern rivers to their headwaters. This coming summer Dennis will be working as a commercial pilot, so I hope to make the trek with one or two women friends.

Studying the maps, my eyes jog along the Continental Divide, and I find scores of delicate-flowing streams and rivers that are born in mountain provinces such as the Philip Smith, Romanzof, and Davidson mountains. My heart is set on finding an east–west route that would traverse several drainages near the crest of the Brooks Range and cross the divide in at least two places. There are an unlimited number of routes, all with spectacular scenery and great wildlife diversity, so the decision is not an easy one.

After pondering the maps by candlelight, and consulting with Dennis on various choices, I decide on a month-long trip that will have two legs. The first leg will begin at the confluence of Cane Creek and the East Fork of the Chandalar River, follow Cane Creek to its headwaters, then head west across the divide into the upper reaches of the exquisite Marsh Fork of the Canning River.

The second leg of the trip will be more strenuous. The trip will begin along the braided Ivishak River, which lies a couple of drainages west of the Canning River. This route will follow the Ivishak to its headwaters, cross the divide through a 5,700-foot saddle nestled between 7,000- to 8,000-foot rugged mountains, then follow the Wind River drainage south. I have chosen this particular route because the Ivishak–Wind River pass is one of the highest passes along the Continental

Opposite: *Northern primrose in bloom, Arctic National Wildlife Refuge*

Divide, and our course will cut through the heart of the remote Philip Smith Mountains. There is always the chance that no one has ever walked this particular route.

Several months later, it is summer, and Dennis is circling over the Ivishak River, trying to locate a gravel bar on which to land the Super Cub. The surfaces of most of the Ivishak's bars resemble a rough washboard, and can be better referred to as boulder bars. Dennis has a difficult time finding a place smooth enough, without plane-eating rocks, to land safely. He finally discovers an acceptable spot, and soon we are rattling across the rocks. After dropping me off, he heads back to the Marsh Fork of the Canning to pick up Sidney Stephens, a teacher-friend with whom I have just completed the first two-week segment of our trip.

Alone in the broad Ivishak River valley, I listen to the peaceful flowing braided river, and reflect upon the first seventy-five-mile portion of our trip. Sidney and I, with another friend, Mary Clare Andrews, had just hiked up Cane Creek's gentle valley, following the exquisite stream to its headwaters. We passed many upthrusted mountains with unusual pastel, brick red, and coal black layers. Some cone-shaped mountains were folded, twisted, or tilted on their sides, while others had concentric sedimentary layers that spiraled toward the sky. Each bend in the river and foot of elevation gained brought us new mountains with different geologic chapters written on their faces.

For days we walked through undisturbed valleys and along nameless tributaries; we saw no other human footprints. We viewed scores of Dall sheep grazing in the high country or picking their way across talus slopes. One blond grizzly paid us a brief but heart-pounding visit near a campsite surrounded by thickets of soapberry bushes. The bear was scarfing down a breakfast of the bitter red berries when we spotted it. The grizzly evidently caught our scent at about the same time we saw it. Startled, the bear took off through the bushes as though in a steeplechase. Within a minute the galloping bear vanished beyond the sloping crest of tundra. All of us were relieved that the grizzly had left our camp, but at the same time we regretted that it didn't stay longer within binocular range.

The route along Cane Creek and the Marsh Fork also offered a diversity of habitats for a number of birds: arctic warblers and redpolls in the willow bushes; semipalmated plovers and sandpipers on the gravel bars; upland plovers and jaegers on the tussocks; and golden eagles and rough-legged hawks soaring above us. Sidney had been thrilled when she spotted a flock of northern wheatears darting among the rocks just below the divide. She had always wanted to see this mountain-loving bird that breeds in northern Alaska then marathons back to Africa to spend the winter.

I'm brought back to the present when the puttering Super Cub breaks the sound of the Ivishak. Today the river sounds like a steady wind blowing through a pine forest. Soon the broad-winged Cub glides down to the bar like a frigate bird, then bounces off the bar a couple of times before rolling out.

Dennis is anxious to try his luck at fishing for arctic char. He unearths his pole from the back of the Cub and heads over to what look like a couple of deep blue holes at the base of some limestone cliffs. Each summer thousands of the sea-running arctic char migrate up many northern rivers within the refuge to spawn and overwinter. Rivers such as the Ivishak, Hulahula, and Kongakut contain many freshwater springs that provide ideal overwintering habitat for char. The abundant arctic char are a major subsistence resource for local Kaktovik residents, who catch the fish in both summer and winter.

Instead of pulling out our poles, Sidney and I are more interested in eating some of the fresh supplies Dennis has brought. We decide on some avocado, onion, and cheese sandwiches for starters.

"Char!" Dennis shouts as I put the sandwich together. Sure enough, he has caught a good-size char on his first cast.

"Another one!" he yells to us a few minutes later.

With that, I grab my fishing pole and lures and head over to join him. Within a half-hour we have caught seven beautiful silvery pink arctic char. Dennis still has several flights to make from Arctic Village, so Sidney and I send him off with all seven fish, figuring there are more to come. Arctic char are always a welcome surprise for Arctic Village residents, since this particular seafaring cousin of the Dolly Varden is not available in the Arctic Village area. Dennis plans to share the catch.

Soon Dennis takes off and tips the Cub's wings a few times, waving goodbye to the only two human specks of life along this far-reaching northern river. We, indeed, feel isolated. The nearest settlement is Arctic Village, some seventy miles to the southeast. The closest city is Fairbanks, almost three hundred miles due south. If we were to hop into a boat and float about fifty miles down the Ivishak to its confluence with the Sagavanirktok River, we would reach our first sign of civilization: the trans-Alaska pipeline. Following the Sagavanirktok and the silver tube north for another fifty miles, we would eventually reach the Prudhoe Bay oil fields.

Sidney and I decide to walk upriver a couple of miles to find a good campsite and another fishing spot. After a mile of walking on gravel bars, we find another fairly deep, crystal-clear fishing hole with not one visible char. I cast a few times, and there are no nibbles. What if Dennis took off with our only chance of fresh fish, and we end up with another freeze-dried dinner?

Fortunately, that's not the case. Within another mile we discover a deep blue, bottomless hole that looks promising. The pool is bordered by a series of prominent cliffs that apparently have fought the river's erosional force. Springwater gushes out of the upper portion of one cliff, showering the rocks with a fine mist; several birds flit through the water's spray. One lower band of rock is thickly covered with green moss, and for a moment I'm reminded of some of those unusual green-covered pinnacles on the island of Kauai.

After a couple of casts, I catch a three- to four-pounder that will provide plenty for dinner and sandwiches. Now that it is mid-August, most of the insects have vanished, and the char seem to be gobbling my lures. Sidney has packed a fly rod, but she decides to wait until tomorrow to try her hand at the art of fly casting.

That evening we eat a delicious meal of pink char and bannock beside a driftwood fire by the river. For a time the northern sky is painted with sunset colors nearly the same hue as our pink dinner.

The next morning we decide it's too beautiful to leave camp. It's warm, probably in the upper 60s, and the char hole looks inviting for a swim. Sidney spends much of the morning experimenting with the new fly rod, her line sweeping back and forth in great arcs from the gravel bar to the pool. Sometimes she snags a scrubby willow bush on the back of her sweater, and I hear an "Ahhhhh!" of frustration. Most of the time she looks like a wayward music conductor following the tempo of some unconventional piece, arms and hands waving all over the place. Although she doesn't get a nibble, she says she loves going through the motions and doesn't care about the catch.

Later in the afternoon, after swimming in killing-cold water, washing our rank clothes, and catching another whopper for dinner, we head upriver a few miles. As soon as we pack up the tent, black clouds start to consume the sky to the south. Within a mile of char-hole camp, the wind starts gusting into our faces, and soon the rain is soaking us. The squall quickly blows by us, and we're in sunshine once again, although the slate sky hanging over the more distant mountains looks threatening.

We reach a large, mile-wide, nameless tributary that merges with the Ivishak from the west. The tributary has an immense alluvial fan of gravel and boulders that yawns into the Ivishak, yet only a couple of small creeks snake through the clutter of rocks. The broad Ivishak valley is impressively open here, particularly with such a broad-flowing tributary cutting through the 4,000- to 6,000-foot gray mountains that flank the river. Across the floor of the two-mile wide Ivishak valley, I count at least ten channels weaving through an enormous floodplain of rocks.

On the other side of the tributary, we set up camp, since the weather is looking more ominous to the south and slowly moving our way. We erect the tent and cover our packs while thunder rumbles in the distance and a dark wave of clouds moves down the valley like an incoming tide. Each valley to the south gradually becomes

American golden-plover (Photo by Hugh Rose)

Caribou on the coastal plain near Beaufort Lagoon (Photo by Ken Whitten)

Tundra swans
(Photo by Hugh Rose)

Great semipalmated plover
(Photo by Hugh Rose)

Fall colors on the tundra (Photo by Debbie S. Miller)

Coastal plain in July (Photo by Dennis C. Miller)

Porcupine herd caribou near the southern edge of the Arctic National Wildlife Refuge's coastal plain (Photo by Ken Whitten)

Pacific loon
(Photo by Hugh Rose)

Coastal plain with braided Sadlerochit River (Photo by Debbie S. Miller)

Robin discovers a caribou antler. (Photo by Dennis C. Miller)

Nameless lake near Hulahula River (Photo by Dennis C. Miller)

Tanned caribou hides hanging on a clothesline in Arctic Village (Photo by Dennis C. Miller)

Debbie backpacking above the Sadlerochit River with Mount Michelson in the distance (Photo by Anne Caulfield)

Polar bear (Photo by Hugh Rose)

sheathed by veils of rain, shadowed by iron-gray clouds, and rapped with thunder. Closer and closer the storm moves toward us, with the constant rumbling of thunder and intermittent jagged bolts of lightning, yet bright sunshine still bathes us as we are a part of that momentary calm before the storm.

Watching thunderstorms move down a valley such as this is an awesome experience. You can actually study and enjoy the movements of the storm for a time without being caught up in it. Instead of struggling to keep dry, or hiding within the shelter of a tent, you can just observe one of nature's wonders in motion. You can soak in the power of the storm and watch the surrounding mountains and valleys get drenched, without getting wet yourself.

Then the wind picks up, and the first raindrops from the periphery of the storm blow into our faces. Sidney and I throw our few needed belongings into the tent and continue to watch the swirling dark clouds and lightning streak toward the mountaintops on the opposite side of the river. The thunder now shakes us. We can hear, as well as see, thick curtains of rain closing in upon us.

"It sounds as if it's ripping the sky apart," Sidney says of the thunderous din.

Then we notice a bull caribou trotting proudly down a gravel bar in the middle of the river. Concentrating on the storm's show, we had almost missed him. The bull is running north, away from the blow. As he passes our campsite, an unusually loud belt of thunder stops the caribou in his tracks.

Startled, the bull first looks behind him at the oncoming storm, then gives our bright orange tent a puzzled look, as though the thunder and tent are somehow connected. The loner, with his new set of ornate antlers, continues to trot north.

Within seconds the rain reaches us, and we crawl into the tent. Soon hailstones are bombarding the flapping tent fly, and we appear to be in the thick of the storm. I look south out the tent's exit hole and can see a band of sunlight behind the fast-moving thunderstorm. This storm is grasping the earth in its fury of swirling clouds, wind, and thunder; then it releases its grip, suddenly and deliberately, after showering and pelting the ground with rain and hail. We watch the storm continue to move north, gripping and pounding the valley, until the thunder is muffled in the distant mountains.

Soon we are in sunshine again, and the sky is flecked with beautiful poststorm cumulus clouds. Our day ends with another delicious dinner with curried rice, and a spectacular sunset of apricot and pink-washed clouds.

❖

A few days later we have moved several miles upstream, just west of Porcupine Lake, which lies between the Marsh Fork of the Canning River and the Ivishak drainage. We hope to scale a mountain that will give us a good view of the mile-long lake, named by USGS geologists who discovered many porcupines in the area in 1948.

The closer we get to the backbone of the Philip Smith Mountains, the narrower the Ivishak becomes. Along one section of the river a band of cliffs rises some thirty to forty feet above the main channel. One rock formation has what appears to be a striking band of quartz, curving like a sidewinder through several layers of sedimentary rock. Many of the sedimentary layers are twisted and folded, thrusted and squeezed, in almost every direction by some unimaginable geologic force.

We are camped near a large patch of willows and are surrounded by dense thickets of soapberries and an occasional blueberry bush on which there are ripe berries. Bears and berries go hand in hand this time of year, so, sure enough, we find fresh grizzly signs all over. The bears in this region seem to be most fond of the soapberries, we surmise, since we discover dung containing piles of berries all around our camp. The bears seem to be inhaling the berries whole rather than chewing them.

I begin to worry about the smell of char juice on our clothing and on the pot bags. Even though North Slope grizzlies are generally not considered fish eaters, the smell could certainly attract them. I carry our heavy 30.06 rifle for emergency protection, but pray I will never have to use it. As a precaution, I nervously play my harmonica when we walk through the willows and other areas of poor visibility, figuring that my novice tunes will scare off just about anything.

Male grizzlies on the north side of the Brooks Range weigh an average of 350 to 400 pounds, about one-third the size of their giant cousin, the Kodiak brown bear. Although much smaller, these northern grizzlies are not to be treated with any less respect. Most of the bears we've encountered within the refuge have run away at the first hint of our scent. But there are bears with unpredictable personalities, and they may be more curious or aggressive. One thing is certain, I did not want to take any grizzly by surprise.

One sultry afternoon we decide to climb up a lookout mountain on the east side of the river that we hope will offer a good view of the Porcupine Lake region and the upper Ivishak we have yet to explore. It's a relatively easy hike up the gentle ridgeline. When we reach the top, there is just room enough for the two of us to sit and gaze at the impressive panoramic view. We first discover that others have used this summit for a lookout. The craggy mountaintop is peppered with sheep pellets, and we also find a few owl pellets, some tiny vole teeth, and scattered white feathers. An old bleached caribou antler lies several yards below us on the other side of the mountain.

Although many of the highest peaks along the divide are obscured, there is much scenery to view at the lower elevations. Looking north, we see the silver-gray Ivishak bending west toward our old thunderstorm campsite. To the south, we face a wall of some of the most remote mountains in North America, mountains that we would soon cross over. In the somber light, most of the mountains are the same uniform dusty gray, although they are highlighted by magnificent erosional patterns. Some mountains are streaked with chutes and gullies that curve around barren-looking pinnacles and jagged outcroppings that erupt from the steep slopes.

The tundra slopes still have a hint of green, although we can see the beginning of autumn golds and reds that will soon completely paint this northern landscape. On the west bank of the Ivishak we spot a large group of Dall rams grazing on a tundra bench a few hundred feet above the river. To the east, we can barely see the edge of Porcupine Lake because of a low-lying ridge that blocks our view. Beyond the hidden lake we can see the undulating slopes of tundra falling gently toward the Marsh Fork of the Canning River, some fifteen to twenty miles away. All of this, and not a sign of human activity. Since Dennis left us, we have not seen or heard an airplane overhead.

Perhaps at this very moment, somewhere in the halls of Congress, Representative Morris Udall and his many supporters are fighting to pass landmark legislation known as H.R. 39, or the so-called d-2 bill. D-2 refers to Section 17 (d)(2) of the 1971 Alaska Native Claims Settlement Act (ANCSA). In addition to distributing a cash settlement and millions of acres to Alaska Natives, ANCSA directed the Secretary of Interior to withdraw up to 80 million acres of Alaska's unreserved public lands deemed suitable for national parks, forests, wildlife refuges, and wild and scenic rivers.

With the passage of ANCSA, subsequent d-2 legislation, passed in 1980, would ultimately establish the most extensive park and refuge system ever enacted by Congress. The Alaska National Interest Lands Conservation Act (ANILCA), in addition to setting aside 97 million new acres of national park and wildlife refuge lands, would designate 56.4 million acres as wilderness and add twenty-six exquisite Alaskan rivers to the National Wild and Scenic River system. Portions of the Ivishak, Wind, and Shiinjik rivers, in the refuge, would be on that list.

Sidney and I are walking through country that is outside the original Arctic National Wildlife Range. New boundaries are being fought over in Washington, boundaries that will determine whether this land beneath our feet remains under the control of the Bureau of Land Management, or whether it will be managed as a refuge under USFWS. Thanks to those dedicated to protecting wild areas for future generations, to those who beat on doors in Washington, and to the many conservationists and grassroots organizations who lobbied representatives from their home states, ANILCA did become a reality.

The soon-to-be established Arctic National Wildlife Refuge would encompass all that our eyes can see from this summit. A vast portion of the Ivishak drainage, along with other free-flowing river systems stretching westward toward the pipeline corridor and south to the Yukon Flats, would all be part of the new addition. The 19-million-acre Arctic Refuge would be the largest unit within the National Wildlife Refuge system, slightly larger than the Yukon Delta National Wildlife Refuge.

A day later we are exploring the narrower portion of the upper Ivishak, which reveals a new series of geologic formations. For a few miles it is as though we are walking through whiskey bottles lying end to end on a table. In several places the Ivishak narrows for several hundred yards, forming small gorges through a buckskin-colored rock formation. As we enter these winding narrow cuts, we expect that the Ivishak gorge will continue, perhaps forming a deeper canyon as we approach the divide.

To the contrary, it's like walking through the skinny neck of a bottle and entering its more spacious bottom. Each time we walk through a narrow gorge, we reach an opening that funnels us into a gentle open valley. The river once again becomes braided, and there is a prevailing feeling of open space, although we are still shouldered by the Ivishak's mountainous valley walls.

Each narrow passageway brings us new surprises as we enter the more open valley. There are new rock formations, unusual upthrusted mountains, with convoluted sedimentary bands, and always the intricate web of tundra plant life along the river.

In one of these valley rooms, the Ivishak's riverbed is covered with parts of an unusual rock band that I have never seen elsewhere in the eastern Brooks Range. Each hunk, piece, or fragment possesses the same characteristics. The rock is composed largely of an iron-red base and topped with a thin coal-black layer. The black layer is extensively fractured into small polygonal shapes, and quartz, the color of white icing, has filled all the cracks. I would later find out that the reddish portion of the rock is iron oxide and that the Inupiat word for iron oxide is "Ivishak." At one time Eskimos in this region reportedly used iron oxide from this river drainage to make red paint.

We later camp on an ideal bluff that overlooks the valley and is shouldered by impressive 6,000- to 7,000-foot serrated ridges and partially obscured peaks. Along the divide there is one striking dome-shaped, chocolate-colored mountain with a remnant glacier flowing like marshmallow sauce over its crest and down its brow. We call it sundae mountain.

The following day dawns partly cloudy, with an autumn nip in the air and a fairly stiff breeze. Our tent is beaded with the raindrops that pattered on us most of the night. In the last couple of days there has been a dramatic change in color on the tundra. The willows have turned to yellow and deep gold, the blueberry leaves to a subdued violet, and the bearberry leaves to a brilliant crimson.

"It feels like winter is coming," I say to Sidney as the coffee's steam momentarily warms my face.

We're both anxious to get under way just to warm up. With any luck we might cross the divide today. The first few miles beyond camp offer great walking along tundra-covered river terraces. A couple of stretches of tundra are as smooth as a golf course, and that gives us the opportunity to look around at the precipitous mountains and the remnant north-facing glaciers that form the backbone of the Philip Smith Mountains.

After three to four miles, the valley grows more restrictive, tundra slopes fall steeply to the river, and numerous rivulets and cascades tumble down from the high country. Soon we are walking through a narrow canyon and are forced to make numerous river crossings. Although not deep, the water numbs our feet. The tortuous creek continues through the confining slit for a couple of miles, and our pace slows to the speed of a waddling porcupine as the stream gradient increases. Finally we reach a waterfall and a deep pool, and luckily discover a route that takes us above the canyon so that we don't have to backtrack. While I climb up from the river, a small flock of rosy-crowned finches lands next to me and seems to be studying my progress. They chirp away with such ease, while I huff and puff. Yes, it would be nice to have wings right now.

As we near the Ivishak's headwaters, the terrain grows ever more rugged. No longer would we find stretches of tundra for easy hiking. From now on we pick our way across steep talus slopes, and occasionally use both our hands and feet for balance. Although we are only two miles from the hidden pass, we decide to make camp because the weather looks foreboding. Thick, dark clouds are sweeping over the divide, and a mean wind is blowing down the valley. A storm is coming our way.

At 4,200 feet we erect our tent on what may be the only level patch of rocky ground in the immediate area. Within an hour of setting up camp, the storm begins to pound us with sleet and hail, and we quickly realize that this is not a brief thundershower. All evening and through the night we are stuck in our tent, listening to the flapping tent and pounding sleet. At one point I go out in the gusting storm to tighten the guy lines and almost need a flashlight; now that it is the third week of August we are experiencing several hours of darkness.

All through the next day the storm continues, and we occasionally peer out the tent door to watch snow falling about 800 feet above us. I worry about crossing the divide. If the snow continues, it may be a treacherous 1,500-foot ascent to the pass over slippery, rock-cluttered terrain. At this point we have no idea whether our chosen route is feasible. Neither of us has seen the pass from the air, and there could be sheer drop-offs on either side of the divide that wouldn't necessarily show up on our 1:250,000-scale topographic maps.

But this was all part of the adventure of exploring this wild country. I was glad we had not seen our route from an airplane. Wilderness explorer and advocate Bob Marshall once wrote: "One of the greatest values of exploration is in pitting oneself without the aid of machinery against unknown Nature. When you use machinery to get the jump on Nature by making her reveal some of her secrets in advance, it seems to me a little bit like peeping at the end of the book to see how the plot will come out."

If we can cross the divide, we are still twenty-five miles from where Dennis plans to pick us up in three days. If we can't reach the Wind River drainage, we will have to backtrack about forty miles down the Ivishak to the closest known landing area. Although it is wearisome to think that our route may not be feasible, there is a tremendous sense of adventure in not knowing what lies ahead. Perhaps one of the greatest values in experiencing this primeval wilderness is the element of discovery.

On the morning of August 23, after we've been trapped in our tent for thirty-six hours, the skies are blue and clear. Our tent is sheathed in ice and is crisp to the touch. The condensation on our bags is frosted. I did not sleep well through the damp, cold night. The temperature is in the low 30s as we awkwardly crawl out of the tent, stiff as a couple of two-by-fours. The rugged mountains along the divide are beautifully dusted with fresh snow, and although they are only seven thousand to eight thousand feet high, the relief is as impressive as some of the glaciated ranges we've seen at much higher elevations, such as the Alaska Range.

Within the hour the temperature warms a few degrees, and we can see that the snow will melt quickly in the pass area. We break camp and head toward the divide at a fast pace to avoid potential high runoff. Soon the rocky valley slopes narrow into a canyon, and we again make numerous river crossings through knee-deep icy water.

Rounding a sharp bend in the river, we discover an accessible ridge above the canyon that will give us a chance to exit the freezing riverbed and gain a view of what lies ahead. As we near the top of the ridge, we see for the first time the 5,700-foot saddle nestled in the 7,500-foot jagged crestline. With our fifty-pound packs, the walk across the steep, loose talus will be difficult, but we're relieved that the route over the divide looks like a go.

Scanning the pass area, we both notice an unusual bright patch of orange about two hundred feet below the pass. I assume that it is lichens contrasted against the slate-gray ridge. But why are they growing in that one particular spot? We pull out binoculars, and to our amazement we discover that the orange is not lichens. It is the remains of an airplane wreck.

Cautiously we traverse the unstable talus, trying to avoid starting a rockslide. We sometimes hear water trickling well beneath our feet, and occasionally the faint rumbling of rocks shifting beneath us. How long had this plane wreckage been lying in the mountains? Were there any survivors? Why did it crash in this isolated, rugged area?

As we approach the wreckage, I pick up an old duffel bag that has rolled down the mountainside. Written in ink on the canvas is a legible address, and the aluminum zipper still works, so I assume that the plane wreck can't be too old. Inside the bag I find six army sleeping bags.

When we get closer to the wreckage, we realize the plane accident must have happened years ago. Vintage 7-Up and Coke cans lie unopened near the crash site. Rusted debris and crumpled cowling are scattered around what remains of the fuselage. Most of the international-orange, amphibious plane had burned. One engine and wing—with the plane's identification numbers, N720—lie intact on the 40-degree slope.

Sidney spots a metal flight case that was evidently thrown from the plane on impact. Strangely, it rests on a large rock as though someone had placed it there. Enclosed in the case, protected from the elements, are old flight logs. Luckily only the lower right-hand corner of the logs burned and the entries are perfectly legible.

The logs reveal that the plane belonged to the USFWS and had been piloted by a man named Rhode. The entries indicate that on August 21, 1958, almost twenty-one years ago to the day, Rhode took off from Porcupine Lake, approximately twenty-five miles northeast. He departed at 10:10 A.M. and returned to Porcupine Lake at 11:35 A.M. He stayed at the lake for almost two hours and then took off again at 1:26 P.M. That was Rhode's last entry. He gave no destination.

An eerie sensation sweeps through us as wind funnels down from the pass, rattling pieces of metal against the rocks. Who was this man named Rhode, and what was the final destination of this plane? It is obvious that no one could have survived the violent crash. Is it possible that we are the first people to locate these remains? Given that twenty-one years have passed, we guess that someone has likely spotted the wreckage from the air. Nevertheless, we decide to carry the flight case with us and report the plane as soon as we reach Arctic Village.

Walking up to the pass, I keep wondering who had given up their lives to these enduring mountains. Soon we reach the saddle and can't help but linger on the spine of this rugged range, where each drop of water is given its destination. The Wind and Ivishak rivers are only gently flowing trickles here. Narrow veins of water form a miniature delta within the saddle, the birthplace of these two wild, clear-flowing rivers. The Ivishak will eventually merge with the Sagavanirktok River and continue beyond the oil fields into the Beaufort Sea. The south-flowing Wind will meander some seventy-five miles through and beyond the Brooks Range, into spruce forests, and merge with the East Fork of the Chandalar. Its waters eventually flow into the Yukon River and on to the Bering Sea, more than a thousand miles to the southwest.

Looking down the Wind River valley, we don't expect any great difficulty on our descent. The first portion goes easily as we walk down the scree-covered slope. Then we hear the faint rumbling of water below us. I picture Yosemite Falls around the bend, with us stranded on the brink. As the rumbling grows louder, I notice a void in the distance. Soon we reach a thirty-foot drop-off, with the falls spilling below us.

"This vertical plunge doesn't show up on the topo map," I say to Sidney.

Much to our relief, we spot an alternative route skirting the falls. The unstable ground is saturated with water, and we pick our way through many loose rocks. Several times we lose our balance, but neither of us tumbles into the streambed. It takes several hours to cover the two-mile descent to the main river as we continue to encounter more rough terrain and a series of waterfalls.

With only two days remaining, we have to hurry to reach the planned pickup spot. Both of us regret that we don't have more time to enjoy the spectacular scenery along the upper Wind River. We've now spent a full month exploring the refuge, yet, as always, I find that the longer the time I spend in this country, the greater my peace within, and the harder it is to leave. A month is just not long enough. It usually takes a couple of weeks to unwind from the pulse of civilization, to revive the dusty senses, and to become acutely aware of nature's northernmost web of life, with all the subtle sounds and signs that are inherent in this great wilderness.

By the third and fourth weeks, that busy human world has grown more distant and my mind is sharply in tune with its natural surroundings. There is a tremendous sense that I am beginning to feel a part of the country. Or perhaps it's that this wildness has become a part of me, in a moment-to-moment, uninterrupted way. The longer I breathe in the pure air, drink the mountain water, and absorb nature's gifts into my mind, heart, body, and soul, the richer my perspective of life; be it the hardy lichens that have grown for centuries, the gentle eyes of rosy-crowned finches, or the multi-million-year-old rocks that make me feel as insignificant as a speck of dust.

Dennis meets us on schedule on the planned pickup day, and one hour later we arrive in Arctic Village. Dennis uses the community telephone to call the Flight Service Station in Fairbanks to report the plane wreckage and its location. He reaches a longtime employee who is amazed by our discovery. The employee immediately recognizes the N720 identification numbers and tells Dennis that we have solved the twenty-one-year-old mystery of the disappearance of Clarence Rhode; his twenty-two-year-old son, Jack; and USFWS agent Stanley Fredericksen in their twin-engine Grumman Goose. The man informs Dennis that in 1958, Clarence Rhode was Alaska's USFWS regional director.

Within two hours of that telephone conversation, news of the discovery is broadcast on Fairbanks radio stations, and eventually throughout Alaska. The following day, banner headlines on page one of the *Anchorage Times* read: Hikers Find Rhode Plane. Sidney and I are amazed to learn that following Rhode's disappearance, hundreds of air force, government agency, and private aircraft conducted the most extensive rescue search in the history of the Alaska Territory to find Rhode's plane. Day and night, pilots systematically searched for the plane, scanning more than 280,000 square miles—an area larger than California and Oregon combined.

After a month of intensive searching, and following a few false leads, the air force estimated that the cost of the search had grown to a million dollars. During the search the weather had deteriorated, with days of snow, icing conditions, and fog. Flying conditions continued to worsen as winter in the Brooks Range set in. Although the air force and others discontinued their searching efforts at the onset of winter, the USFWS refused to give up. Until early December USFWS planes continued to check out every possibility: lakes that Rhode might have crash-landed upon, trappers who might have seen the party, remote cabins that might have been used for shelter. There was even speculation that perhaps Rhode's plane had been shot down, or escorted by the Russians into Siberia. The State Department contacted Russian officials to see if they had any contact with a plane meeting the description of the Goose.

With sub-zero temperatures and only a few hours of daylight remaining in December, the search was called off for the remainder of the winter. The forty-four-year-old Rhode was an accomplished outdoorsman and pilot, and the Goose had been well-equipped with survival gear. Rhode had also taken many

Tent site at Cane Creek in the Brooks Range

trips to the Arctic, a place he dearly loved and knew well. Friends and family refused to give up hope through the winter of 1958-59, having great faith that the party could survive the winter. After a full year had passed, there were still no clues. The missing men were declared dead.

On August 20, 1958, Rhode had left Fairbanks with his son and Fredericksen, en route to the then-proposed Arctic National Wildlife Range. They intended to survey Dall sheep in the Porcupine Lake area and to check out hunting parties in the vicinity. They took 200 gallons of extra gasoline, which was stored in five-gallon cans in the large nose of the Goose. Rhode planned to make fuel caches in a few locations for future use.

Rhode made his last radio contact from Porcupine Lake on the afternoon of August 20. Later that day the party flew to Peters and Schrader lakes, some fifty miles to the northeast. There they visited with an International Geophysical Year scientific party based at Peters Lake. Those scientists were the last people to see Rhode and his party before they disappeared. The flight logs indicate that after Rhode left Peters Lake, he returned to Porcupine Lake.

Dave Spencer, who was then Alaska's refuge supervisor in charge of fifteen other earlier-established refuges, recalls that Rhode planned to return to Fairbanks by August 23, to meet Ira Gabrielson and Clinton "Pink" Gutermuth, then president and vice-president of the Wildlife Management Institute in Washington, D.C. Gabrielson and Gutermuth had hoped to join Rhode and receive a tour of the proposed Arctic National Wildlife Range. Spencer remembers the influential Rhode as being "politically active on a national level" in pushing for the proposed Arctic National Wildlife Range. Rhode was considered the strongest advocate within the Department of Interior to press for the arctic range.

At the time of his death, Rhode was a prominent figure in the field of wildlife conservation. Since the mid-1930s, Rhode had worked extensively as a game warden in Alaska, traveling by horseback in the early days and by plane after he learned how to fly in 1939. He was appointed regional director of the USFWS in 1947, and soon built up the service's fleet of airplanes, which were equipped with the most advanced post–World War II aircraft radio system available. The aircraft radio system helped USFWS effectively manage fish and wildlife resources and nab violators. Dedicated to work in his field, Rhode also helped to establish a department of wildlife management, and a cooperative wildlife research unit at the University of Alaska in Fairbanks.

By no means was Rhode a typical administrative bureaucrat. Instead of being a desk man, Rhode liked to work in the field. During the 1950s, he flew many parties into the proposed arctic range. Rhode's surviving son, Jim, recalls that his father believed that the best way to convince people that the arctic range should be established was by taking individuals to see the area. Rhode was often frustrated with his effort to push for the range because many felt there was no sense of urgency to protect the arctic region; its very vastness and wildness, people seemed to believe, protected it from development. "His philosophy was that if people see it, they will be moved by it and understand the importance of preserving the area," Jim later told me.

Jim remembered that his father had a deep passion for the Arctic because it was the most untouched region in Alaska, and that he believed the proposed arctic range was necessary because one day the pressures of population and economic development would encroach upon that precious region the way they have on so many other areas in the continental United States. The pressure of oil development within Alaska was just beginning on the already-established Kenai Moose Range, south of Anchorage. Rhode feared that such development in southern Alaska also might threaten another important proposed range in the Yukon-Kuskoquim region of southwestern Alaska. At the time of his death, there was no immediate threat of development within the proposed arctic range.

In a letter dated November 8, 1957, Rhode wrote Olaus Murie, president of The Wilderness Society and leading spokesman for the conservation drive to establish the arctic range. In his letter Rhode reported to Murie that progress was being made on setting up the arctic range and that the USFWS had received many endorsements from various places, largely thanks to Murie's influential work. Rhode went on, however, to express his fears concerning oil development within the Kenai Moose Range and the proposed Kuskoquim range (later renamed the Clarence Rhode National Wildlife Range, and still later the Yukon-Delta National Wildlife Refuge). He suggested that Murie might bring some of these issues to the attention of Secretary of Interior Fred Seaton. Part of Rhode's letter to Murie reads:

> This all-important nesting area [the Kuskoquim range] has been "in the mill" for so many years we are getting discouraged. It means so much, as you know, to the wildfowl picture generally and particularly the entire west coast. If oil explorations start out there without safeguards and a film is spilled into those connecting rivers and sloughs, we shudder to think of the consequences. . . .
>
> There is much pressure in Anchorage, backed by the Chamber of Commerce and oil interests, to convince everyone oil exploration and development will not harm moose habitat in any way and might even enhance it on the Kenai Moose Range. Some of the proposals call for a road network in a grid fashion

> every quarter mile. I cannot agree that would be helpful in maintenance of the type of moose habitat which appeals to me but it is difficult to convince these hungry promoters. It even appeals to some moose hunters who feel they would have no difficulty with such a network, of killing moose where they could back up the car to load them. They will twist statements referring to adequate cropping for range protection, and roads for fire protection, into statements that make such proposals beneficial for moose and man.
> We hope to be able to hold at least half the range intact and omit road building in the remainder.

Tragically, Rhode did not live to see that the Kuskoquim and arctic ranges would become established, nor did he see that more than half of the Kenai Moose Range would be left intact, and that oil development and road building would be limited. Also, Rhode died long before the Prudhoe Bay oil discovery and the subsequent development on the North Slope. As he had feared, the hungry developers would indeed inch their way toward the Arctic and encroach upon America's northernmost wilderness. If Rhode were living today, how disheartened he would be to see his former employer, the Department of Interior, recommend to Congress that oil development occur within the nation's last great stretch of arctic wilderness, within the proposed wildlife range he so deeply cherished.

The following week I'm asked to direct an investigative team to Rhode's crash site. It is late August, and the fall weather is wretched for flying through the Brooks Range. Jon Osgood, from the National Transportation Safety Board; Dick Hemmen, a state fish and wildlife officer representing the coroner's office; Jim Rhode; and I fly by helicopter from Arctic Village to reach the site.

The pilot makes several attempts to reach the Ivishak drainage, following different valleys, searching for clear low-lying passes where we can cross the divide, but the snowstorm makes it impossible to press on. Jim pensively sits in the helicopter, flooded with emotion. For so many years he, his sister Sally, his deceased mother, and other relatives and friends, had pondered and grieved over the trio's mysterious disappearance. Now that his father and brother's cause of death had become known, Jim was mourning the loss and painfully reliving the tragedy of twenty-one years ago.

On the second day the cloud layer lifts, and we reach the snow-covered Ivishak valley. With four inches of fresh snow, I can barely make out the crash site. Because of the elevation of the wreckage, the remote nature of this winding, narrow portion of the upper Ivishak, and the fact that the pass area is covered with snow for as much as ten months of the year, it is easy to understand why the search efforts had failed.

The helicopter has flown us in groups of two up to the pass where Sidney and I had stood only a week ago. With temperatures in the 30s, we carefully climb down across slippery, loose rocks to the crash site and spend more than four hours investigating the wreckage. We fan out across the talus, each of us digging through snow to uncover fragments of the Goose that would give us good clues as to how and why the plane had crashed.

A few articles, like Rhode's weathered flight bag, had been thrown from the plane upon impact, escaping fire damage. Inside the flight bag we find many USGS topographical maps, World Aeronautical Charts, and old USFWS forms. The freezing temperatures have prevented deterioration; the maps are damp, but readable. The 1950s vintage topo maps are rudimentary, showing little detail. Instead of the 200-foot contour lines we have today, these old maps were scaled in 1,000-foot contours. Ave Thayer, one of the pilots who searched for the Goose, later told me that pilots in Rhode's day "relied on maps for direction and general topography only." With the lack of navigational aids, detailed maps, and modern radio communication equipment, flying in the 1950s presented challenges unknown today.

Continuing to brush away snow, we find many five-gallon gas cans that had been flattened like pancakes upon impact. When the nose hit the mountainside, the stored gas cans exploded and most of the fuselage burned beyond recognition. Death would have been instantaneous. There are very few human remains because fire and time have eliminated them. But there are enough mechanical clues for Osgood to draw some preliminary conclusions as to the fate of the Goose.

Osgood says there is overwhelming evidence, because of the contour of the twisted and curled propellers, that the engines were under power at the time of the impact. The throttles were in a forward position. Osgood is convinced that, for unknown reasons, the Goose made a left-hand turn as it neared the pass. As the plane ran out of turning space, the left engine and nose hit the mountainside.

Because of the nature of the winding Ivishak, Rhode could not have seen the pass area until he rounded the final bend of the canyonlike valley, about one mile from the pass. That meant he had only a minute or less to make a decision: to either go over the pass or turn around.

Because of the conflicting weather reports at the time, there will always be unknowns as to the possibility of poor visibility, icing conditions, or strong downdrafts. A patchwork of varied weather systems can be distributed throughout the Brooks Range on any given day. Was Rhode heading home to Fairbanks? Did he mistakenly fly up the wrong valley? Much remains a mystery.

Flying back to Arctic Village through the rugged eastern Brooks Range, I regret that I would never have the chance to know this farsighted man named Clarence Rhode. He died on the divide of his beloved mountains on the eve of what would become the national environmental movement of the 1960s. His life ended at the very time the battle began to establish this northeastern corner of Alaska as a wildlife range. Many other individuals and groups would successfully carry on that fight. Those who knew of Rhode would remember his lifelong efforts in the field of wildlife conservation, his love for the Arctic, and his influential role in helping to set aside the Arctic National Wildlife Range.

9

The Dream and the Fight

LONG BEFORE CLARENCE RHODE'S mysterious disappearance, the dream of preserving some of Alaska's northern wilderness was born in the minds of a small number of farsighted individuals. Like Clarence Rhode, these visionary leaders had the fortunate opportunity to experience the magnificent beauty, solitude, and wildlife diversity of the Arctic through their professional work.

Olaus Murie, a prominent wildlife biologist and conservationist, was the first scientist to explore the Brooks Range and to play an influential role in the establishment of the Arctic National Wildlife Range. Between 1922 and 1926, while he was employed by the Bureau of Biological Survey, Dr. Murie made several wildlife-study trips into the Brooks Range. Traveling by dog team or boat, and often accompanied by his wife, Margaret, Dr. Murie explored many river drainages, such as the Koyukuk, Alatna, Chandalar, Porcupine, and Old Crow. He carefully recorded his observations of flora and fauna and drew detailed illustrations that would later become widely acclaimed.

In the summer of 1926, Olaus, Margaret, and their nine-month-old son traveled by boat down the Tanana, up the Yukon and Porcupine rivers, and on to the headwaters of the Old Crow River in the northern Yukon Territory. The Muries, accompanied by Jesse Rust of Fairbanks, had no idea that the exquisite Porcupine River would one day become a protected river system within the Yukon Flats and Arctic National Wildlife Refuge systems.

The upper section of the cliff-bound Porcupine River now lies within the new addition to the Arctic National Wildlife Refuge, and is home to the endangered American peregrine falcon and other raptors. After traveling several hundred miles along the silted Tanana and Yukon rivers, Margaret Murie described the Porcupine in her classic of adventure, *Two in the Far North*:

> We no longer heard the glacial silt of the Yukon passing the sides of the boat, sizzling like something frying in a pan. For the Porcupine is a clear stream, and that means a great deal. First, it is beautiful; the trees and vegetation go on down

> into the water and the rocks beneath lend color to the stream, so that the whole world here feels more crystalline and sparkling. Second, after weeks of trying to "settle" water in kettles by adding mustard, with no success, and of mixing the baby's powdered milk and food with, and washing his clothes in, the "gravy" of the Tanana and the Yukon, the clear water was a treasure.

During these early trips, Dr. Murie focused on his biological studies of northern Alaska, such as a geese-banding project in the case of the 1926 river trip. Some thirty years later, the Muries returned to northern Alaska for a different purpose. As president of The Wilderness Society, Dr. Murie traveled to the Far North in his work to help establish the Arctic National Wildlife Range.

The vision of preserving a large wilderness area in northern Alaska was first voiced by Robert Marshall, a forester, conservationist, and explorer who hiked through much of the central Brooks Range between 1929 and 1939. Marshall thrived on the isolation and scenic beauty of America's northernmost mountains and feared that one day civilization would encroach upon the remote region unless a sizable area was protected from human exploits. Marshall drew public attention to the area through his vivid writings about the spectacular country, and about the few Eskimos and miners who had settled there. In 1933, Marshall's classic book *Arctic Village*, was published. This book was based on Marshall's experiences and explorations in the central Brooks Range while he lived in the small settlement of Wiseman (not to be confused with the present-day Arctic Village).

Marshall was deeply impressed by the sense of isolation in the Arctic, and he believed that exploration was perhaps the greatest aesthetic experience a human being could know. Marshall's appreciation of the value of exploration is described in a 1934 unpublished review, where he wrote:

> There is something glorious in traveling beyond the ends of the earth, in living in a different world which men have not discovered, in cutting loose from the bonds of world-wide civilization. Such life holds a joy and an exhilaration which most explorers today cannot understand, with their radios and aeroplanes which make the remotest corners of the world just a few days or even hours away in distance. Modern mechanical ingenuity has brought many good things to the world, but in the long list of high values which it has ruined, one of the greatest is the value of isolation.

Moved by the spirit of wilderness, the thirty-four-year-old Marshall called together other conservationists and formed The Wilderness Society in 1935. The Wilderness Society would ultimately play a major role in the establishment of the Arctic National Wildlife Range. During the 1930s Margaret Murie remembers when she and Olaus had their first meeting with Bob Marshall in Washington, D.C. She recalls that Olaus and Bob immediately struck up a lasting friendship, and they were soon discussing Marshall's idea of setting aside a large area in the Arctic as a wilderness preserve. Bob Marshall was "full of enthusiasm and eagerness," Margaret said, and he greatly influenced Olaus.

Just as Clarence Rhode had tragically died during the prime of his life, Marshall died at the age of thirty-nine. One year before his death, Marshall offered strong recommendations in a congressional study of Alaska's recreational resources. On behalf of the U.S. Forest Service, Marshall petitioned in 1938:

> Because the unique recreational value of Alaska lies in its frontier character, it would seem desirable to establish a really sizable area, free from roads and industries, where frontier conditions will be preserved. . . . I would like to recommend that all of Alaska north of the Yukon River, with the exception of a small area immediately adjacent to Nome, should be zoned as a region where the federal government will contribute no funds for road building and permit no leases for industrial development. . . .
>
> Alaska is unique among all recreational areas belonging to the United States because Alaska is yet largely a wilderness. In the name of a balanced use of American resources, let's keep northern Alaska largely a wilderness!

Marshall's plea for preserving a sizable area of northern Alaska for its wilderness and recreational values lay quietly during the 1940s. It was a time of war, not a time to ponder wilderness classification of Alaska's northern lands. In 1943, Public Land Order No. 82 proclaimed that all of Alaska's lands north of the crest of the Brooks Range, and the minerals contained within that vast region, be reserved for use in connection with World War II, under the jurisdiction of the Department of Interior.

This sweeping withdrawal, about the size of South Dakota, included the 23-million-acre National Petroleum Reserve—Alaska (NPRA), earlier established by executive order in 1923. This vast national defense reserve consumes more than half of Alaska's North Slope. The withdrawal also included about 20 million acres of the central portions of the northern Brooks Range and North Slope, of which several million acres would later be leased to the oil industry. The Prudhoe Bay oil fields, then a sleeping giant, lay within this future leased area. Finally, the military withdrawal included approximately 5 million acres of lands in the northeastern corner of Alaska that would later become a part of the 8.9-million-acre Arctic National Wildlife Range.

After the war ended, the federal government took a greater interest in studying lands within Alaska that might be suitable for recreational purposes. In 1949, the National Park Service initiated an Alaska Recreational Survey, under the leadership of George L. Collins. It was Collins's responsibility to become acquainted with the Alaska Territory, work with Alaskan authorities to develop a park and recreation program for the territory, and identify areas within Alaska that deserved protection.

In 1950, George Collins traveled across Alaska, met with territorial governor Frank Heintzleman and other officials, and visited with many local people in bush communities. During this same time, Dr. Murie was advised by Joe Flakne, secretary of the Arctic Institute of North America, that he should get in touch with Collins and others involved with the Alaska Recreational Survey to advise them of his concerns for wilderness preservation in the Arctic. Dr. Murie was serving as president of The Wilderness Society at the time, and he began communicating with Collins and others.

In the early 1950s Collins worked closely with field biologist–pilot Lowell Sumner. Sumner flew many hours of wildlife surveys across Alaska, and he contributed much information to the recreational survey. In 1951, Sumner accompanied Clarence Rhode on the USFWS annual wildlife census and inspection by air. Collins, now in his mid-eighties, recalled in a letter to me that Rhode flew Sumner to many places in the Arctic so that he could get the best possible overview, knowing that Collins and Sumner would soon conduct their recreational survey of the area.

Since the navy was in the process of exploring for oil and gas within the NPRA, John C. Reed, a USGS senior official, advised Collins that the National Park Service should concentrate its arctic survey efforts in the northeastern corner of Alaska, steering clear of potential oil and gas development conflicts. Collins clearly remembers that Dr. Reed told him to "stay east of the Canning River, and you'll be all right." Dr. Reed had also traveled extensively through northern Alaska and recognized that the northeastern corner had the ideal characteristics for a future park. In a November 1985 letter to me, Collins wrote:

> John Reed, who knew the NPS in Alaska very well, told us in effect that after we had seen the country we would pick the high region at the eastern end of the Brooks Range as our prime target. We knew that there were many places that would bear careful scrutiny in years to come because of their importance to prehistory, history, biology, and other natural sciences, but we wanted to locate the one prominent region that was most representative of park values. As John Reed thought, that region was it, and in my opinion still is.

In 1952, Collins and Sumner spent much of the summer within northeastern Alaska. They camped at Schrader Lake with a number of scientists, including A. Starker Leopold and Frank Fraser Darling. On the shores of this exquisite lake, with Mount Chamberlin's glaciated peak rising in the distance, these men discussed the idea of protecting the region as a park, wilderness area, or refuge. All agreed that some action was needed, but it was unclear just which status the area should receive.

However, Collins later recalled his first impressions of the region: "From the time I first saw it, it was the finest national park prospect I had ever seen." The region contained the highest glaciated peaks in arctic North America, a complete spectrum of habitats from the south slope of the Brooks Range to the Arctic Ocean, and a tremendous diversity of wildlife, and it was all virtually undisturbed. Collins and Sumner also recognized that the area provided an important landbase for Inupiat and Athabaskan peoples and for the continuation of their subsistence-oriented cultures.

Drs. Leopold and Darling were also overwhelmed by the beauty and untouched quality of the Arctic. They later co-authored a book entitled *Wildlife in Alaska*, based on four months of wildlife studies throughout Alaska. One chapter focuses on the great opportunity for wildlife conservation in Alaska, particularly in the Arctic, as they saw it during the summer of 1952. They concluded:

> Because Alaska is so newly occupied and so slightly altered, there is time for policies of conservation to be conceived, considered, and put in hand before really severe or irreversible devastation takes place. . . . Much of northern Alaska, in point of fact, has scarcely been altered. There are vast expanses where the fauna and flora are essentially virgin. In short, a rare opportunity is presented to predetermine the course of development and utilization of wildlife resources in an area that is still not far removed from original condition.

Later on that same summer, Collins and Sumner camped in the upper reaches of the Shiinjik River, on the south side of the divide. While camped at Last Lake, they began drawing the boundaries for a proposed arctic preserve. These boundaries, drawn in the field, would eventually be used by Secretary Seaton in establishing the Arctic National Wildlife Range.

The following summer their fieldwork continued on the upper Firth River and Kongakut drainage. They also conducted extensive aerial surveys across northeastern Alaska and the northern Yukon Territory. Their studies concluded that the vast, magnificent region was ideally suited for an Arctic International Park, and in 1953, they recommended that representatives from Canada and the United States work together to preserve the area.

Collins and Sumner laid the groundwork for the establishment of an arctic preserve. Fearing that they would receive considerable opposition from miners to the idea of a federally controlled park, Collins and Sumner suggested that the arctic preserve be opened to mineral entry and leasing. Yet, at the same time, they believed that mineral development was unlikely in such a remote area and that the words of John Reed, "stay east of the Canning, and you'll be all right," would probably hold true.

By 1953, Collins and Sumner were drawing support for an arctic preserve from prominent conservationists such as Olaus Murie, Richard Leonard, Howard

Zahniser, and, Sig Olson. Meetings were held to discuss the idea of formally setting up a wilderness area. In 1954, the Conservation Foundation and the New York Zoological Society announced that they would sponsor an arctic expedition to be led by Dr. Murie. The purpose of the trip was clear. Murie and his party were to study the flora, fauna, and wilderness characteristics of the region. After gathering the facts, and ultimately publishing such information, conservation organizations would begin their drive to establish a wilderness preserve that perhaps would extend across the Canadian border.

In the summer of 1956, Olaus and Margaret Murie and their scientific party departed for the Shiinjik River to set up a base camp on Lobo Lake. Dr. Brina Kessel, a University of Alaska ornithologist, and George Schaller and Robert Krear, two graduate students, accompanied the Muries. Soon after arriving at Lobo Lake, Margaret Murie describes life within the Shiinjik River valley in *Two in the Far North*. The following is from her June 3 diary entry:

> I have watched a band of fifty caribou feeding back and forth on a flat a quarter of a mile away; ptarmigan soaring, and cluck-clucking and giving their ratchety call, all about; tree sparrows so close and unafraid; cliff swallows hurrying by; Wilson snipe and yellowlegs calling; gray-cheeked thrushes singing. The three young scientists are beside themselves with all there is to see and do and record. . . . The only sounds we hear are the sounds of the land itself.

While on the Shiinjik, Supreme Court Justice William O. Douglas and his wife, Mercedes, visited the Murie camp. Justice Douglas was deeply moved by his experience on the Shiinjik and would later influence others, in his soft-spoken, nonpolitical manner, to support the establishment of the Arctic National Wildlife Range. Justice Douglas's book, *My Wilderness*, includes one chapter on his Shiinjik trip. He expressed his emotion for the area when he wrote:

> It was difficult to express my feelings as I stood beside these dark quiet pools, shaded by spruce. They were so beautiful, so exquisite, that they were unreal. They seemed withdrawn from this earth, though a glorious part of it. . . . Here was life in perfect ecological balance. A moose had stopped here to drink. Some water beetles skimmed the surface. Nothing else had seemed to invade this sanctuary. It was indeed a temple in the glades. Never, I believe, had God worked more wondrously than in the creation of this beautiful, delicate alcove in the remoteness of the Sheenjek Valley.

When the Muries returned to Fairbanks in August 1956, they were full of enthusiasm for the proposed arctic preserve. They spent the remainder of the summer

Opposite: *Peters Lake and Mount Chamberlin*

talking to Alaskans and drawing support for setting aside some of Alaska's northern wilderness. First, Dr. Murie spoke to the Tanana Valley Sportsmen's Association, vividly describing what he had seen on the Shiinjik and explaining the concept and values of a wilderness preserve in the years to come. In listening to the tape of Murie's presentation, I heard him speak with gentle persuasiveness. He didn't come on too strong, yet he was convincing with his genuine and direct recollections about his experiences in the Arctic. As Margaret Murie later told me, "Olaus had a natural ability to deal with people."

Dr. Murie emphasized to the sportsmen's group that individuals should have the opportunity to enjoy the quality experience of going into the wilderness "on their own two feet," to a place without roads and hot dog stands. Murie suggested that there should be wild areas that are not accessible to mass recreation and hunting, where a person can go into the backcountry on his own and experience nature's serenity. He noted that democracy is based on the same principle of evolution—diversity—and that those people who crave a true wilderness experience should have the opportunity.

Dr. Murie won overwhelming support for the concept of an arctic preserve from the Tanana Valley Sportsmen's Association, a group, along with many other Alaskans, that was initially opposed to any kind of federal withdrawal in the territory. Alaska was soon to achieve statehood. There were many residents who vehemently opposed the concept of federal land withdrawals, military or otherwise. These independents felt that such withdrawals would retard industrial development in the new state, bring the unnecessary burden of additional red tape from outside regulatory agencies, and infringe upon their rights.

One of the major issues yet to be decided was which federal agency should manage the proposed preserve. In March 1957, the Sierra Club sponsored the Fifth Biennial Wilderness Conference in San Francisco. George Collins, still with the National Park Service, chaired the meeting, which many conservationists and land-use agency representatives attended. Collins recalled that there was much discussion about the proposed arctic preserve, and at the conclusion of the conference it was recommended that the Bureau of Land Management designate Alaska's northeastern region as a "perpetual wilderness."

After the conference, a meeting was called at a suite in the Fairmont Hotel to specifically discuss which federal agency should manage the arctic preserve. Present at the meeting were fifteen to twenty people, including George Collins, Lowell Sumner, the Muries, A. Starker Leopold, Clarence Rhode, and others. Collins and Sumner lobbied hard for the concept of an international park, to be jointly managed by Canadians and the United States. It was felt by others that there would be more opposition to the establishment of a national park than to a wildlife refuge because a park would restrict mineral development and hunting, and the National Park Service generally favored new parks that were more accessible to the public in the interest of mass recreation.

Clarence Rhode strongly favored USFWS administration of the arctic preserve. Based on his many years of experience within the territory, he believed that Alaskans would be most receptive to USFWS jurisdiction. Collins remembers Rhode as a "powerful figure," noting that he had an "inside track with the Secretary of Interior." Rhode had personally flown Secretary Seaton all over the territory, and, according to Collins, Rhode could "sell him on the idea."

At the conclusion of the meeting, it was decided that a proposed Arctic National Wildlife Range, under USFWS administration, had the best chance of winning support within Alaska and on the national level. The international park concept was buried for the time being, but the idea was by no means dead. In later years, after the 1968 discovery of oil at Prudhoe Bay and increased oil exploration in the Canadian Arctic, American and Canadian conservation leaders began a second drive to establish an Arctic International Wildlife Range. Although an international reserve has yet to be designated, the Canadians established the three-million-acre Northern Yukon Park in 1984, and in 1987 the U.S.–Canada Porcupine Caribou Herd Treaty was signed to protect North America's largest international-roaming caribou herd. Efforts to form an arctic international wilderness area continue today.

Dr. Murie felt strongly that support for the Arctic National Wildlife Range should begin at the grassroots level. In 1957, the Muries returned to Alaska and continued to speak to groups across the territory, gathering support for the proposed arctic range. Groups including the Alaska Federation of Women's Clubs; three garden clubs (whose national organization amounted to 300,000 members); the Isaac Walton League, Anchorage chapter; and other sportsmen's and conservation organizations were enthusiastic about Murie's presentations and backed the plan.

In May 1957, the Tanana Valley Sportsmen's Association, with about 350 members, made the first public proposal for the establishment of the Arctic National Wildlife Range. John Buckley, director of the University of Alaska's Cooperative Wildlife Research Unit, played an influential role in drawing local support for the proposal, as well as being a strong advocate on the national level. The proposal was endorsed by the Fairbanks Chamber of Commerce and the *Fairbanks Daily News-Miner*. By the time the Muries left Alaska, they were overwhelmed by the positive feedback voiced locally for the creation of an arctic range.

In addition to Dr. Murie's personal campaign in Alaska, Clarence Rhode continued to seek support for the arctic range through the chain of command within the Department of Interior. Rhode flew the Assistant Secretary of the Interior for Fish and Wildlife, Ross Leffler, throughout northeastern Alaska. Leffler, impressed by the magnificence of the area, returned to Washington, D.C., full of enthusiasm for the proposed range. Soon after, the department began drafting legislation to establish the Arctic National Wildlife Range. Under the department's bill, the arctic range would be opened to leasable mineral entry at the secretary's discretion.

Secretary Seaton also took initial steps to modify Public Land Order No. 82, the 1943 military withdrawal that had consumed 48 million acres of the territory. Seaton announced that the 23-million-acre NPRA would remain a petroleum reserve for defense purposes, but that he would relinquish 20 million withdrawn acres, to be opened for mineral leasing and mining entry under public land laws. A scant 5 million acres of the military withdrawal were proposed as part of the 8.9-million-acre Arctic National Wildlife Range.

Many Alaskans praised Seaton's decision to modify Public Land Order No. 82 in that 20 million acres of land would potentially be opened to future resource development, while the 8.9-million-acre Arctic National Wildlife Range would be preserved for future generations, although still allowing mineral leasing at the discretion of the secretary. There were other Alaskans, however, who opposed the proposal as they feared the range would restrict industrial development. Alaska's Territorial Department of Mines called the proposed arctic range a 9-million-acre playground at the expense of possible industrial development." The Anchorage Chamber of Commerce asserted that the range would "cripple development."

Alaskans were split over the issue. While the mining sector and some Anchorage groups denounced the plan, Fairbanks groups supported the proposal, and such endorsements were often reflected in the local paper. The *Fairbanks Daily News-Miner* published one such strong editorial in the fall of 1957, which read:

> We favor the proposal for the Arctic Wildlife Range. We think the complaint of those opposing it is akin to that of a small boy who has just been given a pie much larger than he can eat but who cries anyway when someone tries to cut a small sliver out of it.
>
> We ask those who would raise strong protest over reserving this comparatively "small sliver" to stop and ponder the fact that the 20,000,000 acres now being made available for development by Secretary Seaton's action comprises an area which exceeds the total land area of five New England states combined.

Between 1956 and 1959, public awareness and support for the proposed arctic range increased on the national front through the work of several conservation organizations, including The Wilderness Society, Sierra Club, Isaac Walton League, Conservation Foundation, and the Wildlife Management Institute. Articles about the proposed arctic range frequently appeared in publications such as the *Sierra Club Bulletin* and the *Living Wilderness*. Marshall, Collins, Sumner, and the Muries had planted the seed for the idea with their early work, and during the late 1950s, national organizations nurtured the concept. Howard Zahniser, executive secretary of The Wilderness Society, and Ira Gabrielson, president of the Wildlife Management Institute, both in Washington, D.C., played key leadership roles in uniting the national conservation organizations.

David Brower, then executive director of the Sierra Club, recalls that he and other conservation activists, such as Richard and Doris Leonard, kept in close communication with George Collins and Lowell Sumner and pressed for the establishment of a national park, and later the arctic range. Brower recalls that through the 1950s, he and other leaders, such as Zahniser and Gabrielson, met annually at the North American Wildlife Conference, where strategies were discussed for protecting the area.

The dream of preserving the northeastern corner of Alaska began to bloom. In May 1959, Senator Warren G. Magnuson, a democrat from the state of Washington, introduced legislation drafted by the Department of Interior to establish the Arctic National Wildlife Range. Hearings on the Senate bill, conducted by the Merchant Marine and Fisheries Subcommittee, were held in Washington, D.C., and in several cities in Alaska. Alaska Senator Robert Bartlett, who strongly opposed the legislation, presided over the hearings. The fight had begun.

The arctic range proposal was the first national conservation issue that brought politicians from the nation's capital to the new, forty-ninth state. The proposal acted as a catalyst for the formation of Alaska's first statewide conservation organization, composed of legitimate Alaskan residents, the Alaska Conservation Society. In its embryo state, the loosely organized society had about a dozen members based in Fairbanks who were accustomed to informal living-room meetings. Longtime Alaskan Ginny Wood, who is still an active conservationist, edited and hand-mimeographed the society's newsletter. Celia Hunter, a past president of The Wilderness Society and now a renowned Alaskan conservationist, teamed up with Wood to help run the new grassroots organization.

"We operated off the kitchen floor. Most of us lived in cabins with no telephones. Once a week we would gather for brown-bag-lunch meetings at the University of Alaska to discuss issues relating to the arctic range, and to prepare for the upcoming hearings," Ginny and Celia recalled.

The conservationist fight to establish the arctic range was slightly different in the late 1950s than it would be today. Although national conservation organizations played key roles in promoting the legislation, the bulk of the testimony in support of the arctic range would be voiced to Congress by local residents at subcommittee hearings in Alaska.

More weight was placed on the voices of Alaskans and on personal campaigning by national conservation leaders. Margaret Murie remembers that when Olaus became president of The Wilderness Society, the society had 956 members. Based in their log home in Moose, Wyoming, Olaus personally carried on a letter-writing campaign, dictating his words to Margaret, who typed them.

In the wake of the 1960s national environmental movement, The Wilderness Society, Sierra Club, and other conservation organizations now have hundreds of thousands of members, with regional or state offices scattered across the country. Today's mass communication network is a far cry from the Muries' personal letter-writing campaign. In 1959, the politicians came to Alaska to hear local residents. Today local Alaskans generally fly to Washington, D.C., if they want to testify before Congress on current conservation issues, or paid lobbyists and professionals from representative organizations carry on the fight.

In Alaska 142 residents from various communities testified before, or submitted letters to, the Senate Merchant Marine and Fisheries Subcommittee in October 1959. All tallied, 73 individuals or group representatives testified in favor of the creation of an Arctic National Wildlife Range, 53 were against the legislation, and 12 were either for or against the proposal provided the bill was modified.

Ginny Wood's testimony captured the spirit of those who believed it necessary to preserve some of Alaska's vast wilderness for future generations. She stated:

> We Alaskans must reconcile our pioneering philosophy of conquer, cut, shoot, plow, mine, and move on, to the realization that the wild country that lies now in Alaska is all there is left under our flag.
>
> Those who see the wildlife range as a threat to their individual rights refuse to face the fact that unless we preserve some of our wild land and wild animals now, the Alaska of the tundra expanses, silent forests, and nameless peaks inhabited only by the caribou, moose, bear, sheep, wolf, and other wilderness creatures can become a myth found only in books, movies, and small boys' imaginations as the Wild West is now. And I regret as much as anyone that the frontier, by its very definition, can only be a transitory thing.

Several scientists brought forth testimony noting the importance of preserving arctic habitat for the great diversity of northern species. University of Alaska mammologist William Pruitt stressed that northern mammals such as the caribou, polar bear, grizzly bear, and wolf require a significantly larger home range than species of more southerly latitudes. In order to protect the home ranges of these northernmost animals, Pruitt suggested that the 9-million-acre arctic range was not optimal in size, but minimal, particularly given the long-distance migratory habits of the caribou.

There were others, however, who saw no value in establishing an arctic range. Wenzel Raith, an electrician, felt that all of Alaska should be left open to the rugged individualist, the gold diggers, and the sourdoughs. "I did not come here to be cradled round with womblike care, or to be mastered about by some paternal bureaucracy," Raith said. "Where would we be today if it weren't for the trailblazers, men like Daniel Boone, James Bowie, Kit Carson, Davy Crockett, and Jim Bridger?" he questioned. And there was Joe Vogler, who believed an arctic range would infringe upon his rights. Vogler, considered a colorful and outspoken Alaskan, said:

> I can't see how Alaska is ever to be developed by denying people the right to use the land that is here. I suffer from claustrophobia—I came here from Texas—there was too many fences, too many No Trespassing signs, too many vested interests. If you wanted to go hunting, you had to belong to a hunting club. And I have enjoyed more liberty in this country, more freedom, with the right to go anywhere I wanted to, and now they're gradually closing it up.

Warren Taylor, attorney and speaker for the Alaska House of Representatives, voiced his opposition to the bill. He, along with Governor William Egan, the state legislature, and the Department of Natural Resources, believed that the proposed range would infringe upon potential resource development, and reduce the state's entitlement to federal highway funds. In the spirit of independence and as residents of a new state, many Alaskans and politicians denounced the idea of any federally controlled lands.

Yet in the face of the state's opposition, the U.S. House of Representatives considered and passed the arctic range legislation in 1960. But things were not so easy on the Senate side, primarily because Alaska's U.S. Senators Bartlett and Gruening blocked passage of the arctic range bill. Fueled by the support of Governor Egan and the state legislature, Bartlett and Gruening refused to compromise on the proposal. In an effort to convert Secretary Seaton, Governor Egan went so far as to propose a wildlife management area within the proposed arctic range boundaries, to be controlled by the state of Alaska. Secretary Seaton did not accept Egan's offer.

In the waning hours of the Eisenhower administration, it was clear that the arctic range legislation would not pass Congress. Seaton realized that he had only a few weeks left in office. Convinced that the arctic range proposal should move forward, he had one last option. On December 6, 1960, Secretary Seaton signed Public Land Order No. 2214, which established the Arctic National Wildlife Range for its wilderness, wildlife, and recreational values, by executive proclamation, and closed the area to mineral entry. At the same time, Seaton revoked Public Land Order No. 82, which opened up some 20 million acres in Alaska to potential resource development. Many would praise Seaton's decision as a noble and farsighted act; others would call it a reasonable compromise; and some Alaskans would denounce it.

On the day that the arctic range was established, Olaus and Margaret Murie were attending hearings in Idaho Falls regarding a proposed dam on the Snake River. They received the news on the following day by telegram from friends in Fairbanks. The telegram said in part that "the American people have received a great gift in the Arctic Wildlife Range." Margaret Murie later wrote to Fairfield Osborn, director of the Conservation Foundation, reflecting upon news of the victory. She wrote: "We both wept—and I think then we began to realize what a long and complicated battle it had been."

After John F. Kennedy was elected president, Governor Egan requested that the arctic range public order be revoked. Secretary Stewart Udall reviewed the matter and rejected Egan's request, giving full support to Seaton's farsighted decision. Senators Bartlett and Gruening, who detested Seaton's action, stubbornly blocked all funding for management of the Arctic National Wildlife Range for eight years. Following Bartlett's death and Gruening's failure to be reelected, funds were finally appropriated for the Arctic National Wildlife Range for the first time in 1969.

What convinced Secretary Seaton to make his final decision of proclaiming the Arctic National Wildlife Range? There were many factors involved in the decision. Jim Rhode recalls that Seaton visited his home after his father had been missing for more than a year. While expressing his condolences to the Rhode family, Seaton vowed to Mrs. Rhode that he would go forward with the arctic range proposal, perhaps because he knew how much Clarence Rhode had loved the Arctic and, in the words of George Collins, perhaps because Rhode himself had "sold him on the idea."

Of course, there were many individuals who played key roles in influencing Secretary Seaton. There was Dr. Murie's campaign, and his impressive slide show on the arctic range that he had personally presented to Secretary Seaton. Conservation leaders in Washington, D.C., pressed hard for a decision before Seaton left office. David Brower recalled that in the final days of the Eisenhower administration, Department of Interior Solicitor Ted Stevens resisted the idea of an administratively established arctic range. Brower said that it was Pink Gutermuth, vice-president of the Wildlife Management Institute, who personally fought such internal resistance. In a statement before the Energy and Natural Resources Committee, Brower noted:

> Day after day as the President's last day grew closer, the proposal was stalled. Then Pink in his own way came into the Secretary's office like a storm out of the Gulf of Alaska starting gently and building into a rage. In final exasperation he said in no uncertain terms, "All I want to say, Mr. Secretary, is are you just sweet-talking us, or are we going to get this damned wildlife range proclaimed by the President." Anyone who knew Pink Gutermuth then would know that it was easier to proclaim a refuge than to face the fury of a Gutermuth gale.

One thing was certain. Seaton's 1960 farsighted decision proclaimed the largest wildlife and wilderness sanctuary in the United States for all Americans. In the coming years, the predictions of Robert Marshall and others would hold true. Civilization and industrial development would gradually creep north into the remote Arctic. The trans-Alaska pipeline and its companion transportation corridor would eventually cut through the heart of the Brooks Range, near where Robert Marshall carried on his early explorations. Giant oil fields would be tapped in and around Prudhoe Bay, pumping two million barrels of oil per day into our nation's supply, and billions of dollars in revenues into Alaska's petroleum-based economy. The isolated Arctic and its diversity of residents would come to know a cancerous web of roads, pipelines, and drilling rigs, and all forms of industrial pollution.

In the coming years, the Arctic National Wildlife Range would stand alone as America's last, great untouched wilderness remaining in the Arctic. Thanks to the vision and work of individuals mentioned in this chapter, and many others, we can be grateful for this rare and northernmost sanctuary. In return we must strive to see that such dedicated past efforts and accomplishments were not done in vain, and that the dream of this perpetual arctic wilderness be realized even in the face of our forever-shrinking planet.

Part Four

OIL VERSUS WILDERNESS: THE FUTURE

Robin looking through a scope at the coastal plain, Arctic National Wildlife Refuge

10

A Week at Prudhoe Bay

A FIERCE POLAR WIND ROCKS our small, unheated trailer as if it were a lightweight skiff at sea. Granite-heavy fog swallows us. The steady gale whistles through broken glass and rattles the flimsy corrugated metal that sheaths this temporary home, once occupied by oil field construction workers.

Hunched over a desk, I write wearing a down parka and wool gloves. Robin, our two-year-old daughter, is asleep on the floor, snug in a heavy down bag on an old mattress penned between the desk and bunk beds. Dennis works on the Super Cub outside, in the lee of an old warehouse, waiting for the fog to lift.

It's a wind-howling, fog-drenching July day within the Prudhoe Bay oil fields. Dennis is here to locate radio-collared caribou via his Super Cub for an Alaska Department of Fish and Game (ADF&G) research project. Robin and I were invited to tag along for the six-day work period. We are based at Service City, one of the largest abandoned oil field construction camps, located about fifteen miles northwest of Deadhorse, in the heart of North Slope oil development. We are about eighty miles west of the Canning River, which forms part of the western boundary of the Arctic National Wildlife Refuge.

During the 1970s oil construction boom, Service City housed several hundred oil workers in long trains of modular units equipped with sleeping, dining, and recreational facilities. Now the impoundment is a ghost town, with empty rows of trailers surrounded by piles of surplus equipment, junk, and garbage. When the oil market fell into a slump a few years ago, the company that operated Service City and held lease to a few hundred acres of state land went belly-up. Although some of its equipment was removed from the premises, an incredible amount of junk remains. The bankrupt company sold one of its three-room housing units, to ADF&G since the agency's field biologists were in need of living quarters. Now, in July 1988, the only people at Service City are a handful of biologists conducting research during a few months of the year.

The fog begins to lift. Outside the muddied window, framed by half-hanging curtains, I can see the tops of two fifty-foot-tall floodlights that at one time illuminated the southern edge of the camp's huge gravel pad. Beyond the poles and a small bluff is a slough of the Kuparuk River. Whitecaps fleck the hundred-yard-wide channel, and a strange half-built steel bridge is visible through the fog. On the other side of the channel is an old 5,000-foot gravel airstrip hemmed in by scores of unoccupied trailers and modular units, stacks of old rusted pipe, tires, crates, and piles of unopened sacks of chemicals and additives used to make drilling mud.

Why the roadless bridge? The steel bridge was originally built to connect the airstrip/storage area to the main camp during the few months of the year when the slough filled with water. Most of the year the ice-covered slough was crossable by vehicles. The company planted two dozen hefty steel bridge footings, some fifteen to twenty feet high. Since the bridge was needed for only a brief period during the year, temporary planks were placed on the footings; the company never really had the need to complete the project. The steel pipe legs are encased in cement and will probably remain strung out across the slough forever.

About a mile beyond the roadless bridge and industrial junkyard shines a series of pipelines that are braced several feet above the tundra. The silver tubes, known as "feeder lines," carry oil from Eileen West End's production wells to one of a dozen "flow stations," scattered throughout the Prudhoe Bay and other neighboring oil fields.

Flow stations, also referred to as "gathering centers," are the workhorses of the oil fields. These production facilities separate the oil, natural gas, and water; pump as much as two million barrels of oil daily into the trans-Alaska pipeline; and reinject natural gas and water into the oil field for greater oil recovery. These twelve stations run on powerful gas turbines that, if combined, would be capable of powering an average size city. Today's wind muffles the roar of the generators. When we first arrived at Service City, the wind was relatively calm and the groan of the turbines was noticeably loud, especially in contrast to our previous two weeks in the nonindustrialized arctic refuge.

As the fog dissipates, one of British Petroleum's (BP) drilling pads comes into view. This pad is one of more than 150 production and exploratory pads flung across the tundra like a scattered deck of cards. Each production pad consumes about thirty acres and generally contains a dozen to seventy production or reinjection wells along with a large two- to three-story manifold building. The boxed-in wells rise up from the tundra in neat little rows. They resemble a series of outhouses when seen from a distance. Behind the hospital-green well houses are rectangular bodies of water known as "reserve pits." A drilling pad generally has two to six reserve pits, each about the size of three or four football fields.

Reserve pits spell trouble for the oil industry. The pits are intended to hold drilling mud wastes and cuttings that theoretically settle in the pits for later removal. Drilling fluids are used by the oil industry for cooling and lubrication during the drilling process. The fluids contain potentially toxic concentrations of heavy metals, hydrocarbons, and other chemical additives. One well uses an average of about ten to fifteen thousand barrels of drilling mud, and more than fifteen hundred wells have been drilled on Alaska's North Slope.

Permafrost and low evaporation rates in the Arctic inhibit the natural loss of reserve pit fluids, unlike pits in more temperate zones. Over the years many of the unlined, poorly designed reserve pits have overflowed or leaked drilling mud contaminants into the surrounding tundra, ponds, and lakes. For several years the state of Alaska permitted the oil industry to discharge waters from reserve pits onto the tundra because the pits commonly overflow from annual snowmelt. As a result of this discharging and seepage, USFWS studies documented higher levels of hydrocarbons, chromium, barium, arsenic, chloride, and sulfates, along with greater turbidity and alkalinity, in sampled ponds located near reserve pits.

Poor water quality has reduced population levels of invertebrates, particularly crustaceans. In some cases ponds are void of certain organisms. Biologists fear that contaminated waters and sediments may adversely affect higher members of the food web, such as the red-necked phalaropes, pectoral and semipalmated sandpipers, dunlins, and old-squaw ducks. These migratory birds feast on pond invertebrates during summer months. Studies on the effects of such contamination on birds have only recently begun.

Hundreds of components are used in the preparation of onshore drilling muds: from harmless clay and walnut shells to toxic hydrocarbons and chromium. Like a good chef, each oil company keeps specific drilling mud ingredients an in-house trade secret. The success of well drilling is dependent largely on the type of drilling mud used. Although the Environmental Protection Agency (EPA) requires that industry disclose elements used in drilling muds for offshore drilling, through what has been described as a loose permitting procedure, EPA does not have such a requirement for onshore drilling. As a result, an environmental field monitor on Alaska's North Slope has only a vague idea of what elements, or quantities of elements, are used in onshore drilling muds. When a monitor looks at a sack of drilling mud components, it's much like looking at a tub of margarine with no specific ingredients or additives listed on the container.

There has been a long-standing dispute over whether drilling muds should be managed and disposed of as bona fide "hazardous waste." Heavy concentrations of some drilling mud components can be toxic to humans and wildlife, yet EPA has declined to classify mud waste as hazardous. This is largely because many of the components of drilling mud are nontoxic, and it would be extremely costly, because of the immense volume, for the oil industry to separate and properly dispose of those elements considered hazardous in high concentrations.

While drilling fluids and wastes associated with oil development currently remain exempt from regulation as hazardous wastes, a recent 1987 Report to Congress by the EPA presents alarming information concerning management of oil development wastes. The study looked at eleven oil and gas production zones scattered across the United States, including Alaska, and assessed the damage and potential danger to human health and the environment from wastes.

The report notes that researchers spent merely three months gathering case information concerning damage to human health, livestock, and the environment in a number of states. Yet even this preliminary report documents numerous cases where ground and surface waters have been contaminated by reserve pits or drilling activities. Cases in the report include: poisoning of livestock and subsequent deaths; milk contamination in dairy cows; contamination of domestic water supplies; soil contamination; loss of crops; complaints of sickness, nausea, and dizziness in humans; and, in the case of Alaska's North Slope, contamination of tundra ponds and loss of aquatic organisms.

The oil industry has explored and developed the Arctic for the last thirty years, disposing of millions of tons of drilling muds in hundreds of seeping reserve pits. These reserve pits are America's toxic reservoirs of the Far North.

Although drilling mud is not regulated by the EPA as a hazardous waste, the state of Alaska has recently begun to monitor reserve pits. The drilling mud leakage problem has forced the oil industry to clean up and redesign the pits under state regulations implemented in 1987. Oil companies are in the process of spending millions of dollars to "close out" their old, unlined seeping reserve pits. The industry is no longer allowed to discharge reserve pit fluids into the wetlands. Some oil companies are building smaller pits, about the size of one football field, and attempting to recycle mud waste in their drilling operations. According to one source, the industry has taken steps to remove the more toxic components, such as chromium, from their drilling muds.

Newer state-regulated pits are supposed to be safer environmentally, as the pits are lined with state-of-the-art material to prevent leakage. Some oil companies are choosing to reinject their wastes below thousands of feet of permafrost, a practice some believe may be a viable method for centuries, or until Alaska's permafrost thaws. But won't this method of disposal of millions of tons of drilling mud only hide a legacy of contaminants for future generations?

Although the industry has been forced to make some waste disposal improvements, disposal techniques in high latitudes are far from being considered reliable or safe. As one environmental field monitor noted, "total containment" of drilling mud fluids is the goal; however, there is no disposal method yet that can be considered tested and foolproof.

Later in the morning, the wind's gale force has diminished enough for Robin and me to take a walk around Service City. We explore the slough, in the lee of its bank, so Robin isn't blown off her feet. Although she is bundled up for the dead of winter, she's delighted to find a sandy beach and shouts "Hawaii!" through the wind's roar.

Along the slough we find all kinds of junk embedded in the sand and mud: coils of twisted wire, steel nails, screws, bolts, empty plastic bottles of antifreeze, sections of pipe, tire rims, Styrofoam, and other industry artifacts. Robin runs ahead, then suddenly stops and says, "Caribou . . . caribou," while pointing at the ground. Sure enough, she has spotted fresh caribou tracks heading across the slough. The tracks somehow look out of place amidst the trashed shoreline.

Robin scrambles up the bank to investigate rows of discarded junk, several hundred yards long that encircle the camp. Stacked on wooden pallets are rolls of sheet metal, boxes of stovepipe, telephone wire, portable toilets, an old washing machine, a freezer, an institutional food warmer, an electric furnace, numerous cans of paints, a rusted water heater, tires, lubricant and antifreeze cans, and many wooden crates. I laugh at one crate, which reads: "1975 Mobil Oil Permafrost Cement. Reusable Container. Return to Anchorage." Would any of this junk be returned anywhere?

Several different companies have owned and operated Service City under the twenty-year-old state lease. Although the state required a small bond, provisions contained in old state leases are weak in terms of cleanup and restoration requirements. Numerous North Slope service companies have gone bankrupt in recent years, each company leaving behind its share of discards. One state official told me that the state's greatest hope for cleanup is that a few junk men either can be hired or will assume the old leases and will salvage the remaining scrap metal and other materials. The state's cost to completely clean up a camp like Service City would be astronomical.

We walk across the table of gravel that leads to the abandoned trailers. Robin picks up pebbles and throws them toward the slough. The oil industry uses an incredible amount of gravel for petroleum development in the Arctic, leaving immense pockmarks along river floodplains. Before erecting any facilities, the industry must construct gravel pads, roughly five feet thick, to protect and insulate the underlying layer of permafrost.

The number of pebbles required for building pads and roads is astounding. One estimate reveals that the Prudhoe Bay field along with four neighboring oil fields have consumed as much as 60 million cubic yards of gravel from different material sites along river streambeds and floodplains. That's the equivalent of 6 million truckloads of gravel. If we placed those 6 million average-size gravel trucks bumper to bumper, there would be a line of gravel stretching more than twenty thousand miles, which is nearly the circumference of the earth.

Gravel removal, both authorized and unauthorized, has caused major adverse impacts on aquatic habitat. The most serious problems include alteration of natural river drainages, erosion, sedimentation, and blockage of fish passage. Such disruption has prevented or entrapped fish from migrating to preferred habitat, and increased fish mortality. As a result, the industry is now removing gravel from sites other than streambeds.

Beyond the junkyard, an open trailer door bangs in the wind. Through a dimly lit passageway, we enter one of the camp's long, vacated trailers. The tunnel of a hallway leads us past numerous run-down, dank rooms. Several of the small rooms have broken door frames, as though a gang had looted the place. On some of the beds there is a pile of musty sheets and blankets, as though a shift of workers were homebound and had left the linens to be laundered.

We walk into one curious room that could have been a time capsule. Flung on an unmade bed is a pair of black-and-white checkered cook's pants. Two pale blue shirts hang in an open closet. Attached to the curtain rod of a broken window is a grease-covered kitchen witch doll. The chef's room. This fellow, by all signs, had apparently dropped his work pants and fled his room at the camp's closing.

A chilling wind moans through the broken window, lightly tossing the pages of two old magazines left on the dresser. They are dated April and May 1985, the year world oil prices plummeted. On the floor next to a small drift of snow is a moldy box of institutional recipes. One recipe card sticks up out of the box: Chili Con Carne for 100. Perhaps that was the last supper for the crew of construction workers.

Farther down the trailer we pass the dining room, with its sagging, vinyl floor and low ceiling. Numerous dining chairs are neatly tucked beneath several round tables. Beyond the dining room is a surprisingly clean bathroom, with rows of sparkling white sinks, each with its new bar of unused soap and fresh towels. Whoever last cleaned the bathroom must have prepared for a crew of workers that never arrived.

We soon find an old recreation room with a moldy pool table; the multicolored balls still lie on the table as though a game were in progress. Past the pool table is a small weight room with rusted barbells on the floor, and an old rowing machine. A cedar-paneled sauna that still looks usable is connected to the exercise room.

As we leave the ghostly camp, I think of Alaska's boom-and-bust history of resource extraction: Russians in search of fur seals, miners hungry for gold, whalers in search of the bowhead, trappers catching furbearers, lumber companies harvesting ancient trees for Japan, fishermen hauling fish to the canneries, and the oil industry pumping out Alaska's petroleum. The levels of extraction of these precious resources have all been determined by the market value. The higher the price, the bigger the boom; the more that is taken, the greater the state revenues and the loss of resources; and in the case of timber, gold, and oil, the more wild habitat lost.

The Prudhoe Bay oil field is the largest oil field in North America. It is the biggest of Alaska's booms, and it leaves the longest industrial trail of development. With the recent fall of world oil prices, Service City has become just a part of that industrial trail—a small part of what someday will become Alaska's biggest bust.

A few mornings later, on the Fourth of July, the wind still howls. I crawl out of my sleeping bag and stumble out of the trailer into a mean wind. I can't face the idea of staggering into the icy blow, half-asleep, to relieve myself, knowing that the wind will bite my exposed skin, spray my pants yellow, so I put it off and head into the kitchen trailer for something hot to drink.

In the kitchen, biologist Walt Smith looks as though he's making firecrackers. On the table are metal darts with small, red pom-poms attached to them. Ray Cameron, another biologist, kneels on the floor next to a kit of syringes, darts, and glass vials. Ray wears surgical gloves and uses a syringe to carefully fill the darts with a potent drug. These biologists are preparing the darts for use in capturing caribou that roam in the area.

Loading darts is serious business. There is no talking in the room, and the concentration level is high. One drop of the potent sleeping drug can kill a human,

yet the same drug puts caribou to sleep only temporarily. I grab a cup of coffee and quickly leave the trailer since it's in an "off limits" mode.

Soon the helicopter arrives, and the work force takes off to capture several caribou, weigh and measure them, take blood samples, and, if necessary, place new radio collars around their necks. In this particular study, researchers plan to monitor the body weight and general condition of collared animals in an effort to see what effect oil field activities may have on caribou access to suitable forage, weight gain or loss, and reproductive success. Researchers in Norway believe that some reindeer have experienced slower summer growth rates when frequently disturbed by roads, traffic, and other human activity. There is substantial evidence that if caribou fail to achieve adequate body weight and condition during the short summer season, cows may not be able to successfully reproduce, and under extreme circumstances some animals may not survive the winter.

Since the mid-1970s, ADF&G biologists and others have conducted extensive studies on the effects of oil development on the Central Arctic caribou herd. Unlike the Porcupine caribou herd, the range of the 18,000-member Central Arctic herd includes the North Slope's oil fields. Central Arctic animals must often detour around obstacles such as pipelines, roads, and other facilities. Cows and calves are particularly sensitive to disturbance during calving and postcalving times and tend to avoid areas of human activity. It is believed that a portion of the Central Arctic herd traditionally gave birth to their calves in the immediate Prudhoe Bay area. Now that industrial zone is no longer used for calving.

The oil industry is quick to point out, however, that the Central Arctic herd has continued to grow. The herd numbered only five thousand animals in the mid-1970s and has continued to increase each year by more than 10 percent. Biologists agree that the herd has increased in size, but they believe there are several reasons for the steady growth rate: there are fewer predators, such as wolves and bears, on the North Slope, as a result, in part, of road access and increased hunting and poaching; the herd has experienced a series of relatively mild winters; and since the Central Arctic herd is relatively small, sufficient habitat remains for calving and foraging despite oil development within its range. This picture may change in the future, however, if the herd size increases and/or additional oil fields are developed.

Biologists point out that if oil development were to occur on the summer range of the Porcupine caribou herd, within the 1002 area of the Arctic

Opposite: *Caribou migrating across the coastal plain*

Refuge, a completely different set of factors would have to be considered. The 180,000-member Porcupine herd is ten times the size of the Central Arctic herd, whereas the Arctic Refuge offers only about one-tenth of the coastal plain habitat available in the Central Arctic. Because the Arctic Refuge coastal plain is in close proximity to the Brooks Range, the Porcupine herd is exposed to more predators, which are abundant in the foothills and mountains. Caribou encountering oil field development on the Arctic Refuge coastal plain might be diverted into those areas of higher predation risk. Perhaps one of the greatest unknowns is how a large aggregation of 40–80,000 caribou, seeking insect relief along the coast, will react to a maze of pipelines, roads, and traffic. No one knows, but experiences at Prudhoe Bay indicate that there is cause for concern.

Later in the morning, Robin and I take a walk out on the tundra, northeast of camp. In the distance the rolling tussock country looks green and clean. I speculate that we will find little garbage on the tundra if we walk above Service City into the prevailing wind. Wrong. I forgot that twenty miles of facilities scattered around the northeast would contribute their share of garbage. It is all over the place.

Every dozen steps or so we stumble upon a piece of plastic, a paper product, or, most frequently, Styrofoam. I spot several of the snow-white cups submerged in tundra ponds, wedged beneath rocks and mud along the slough, or snug beneath a tussock. The ubiquitous cups are sprinkled all over the tundra and stand out like white male ptarmigan on their summer nesting grounds. One biologist reported to me that the white Styrofoam makes aerial surveys of tundra swans more time consuming and costly because it is difficult to distinguish a swan from the white plastic.

"Birdie . . . huggie," Robin says with a smile. She has spotted a pair of Lapland longspurs and, as usual, would love to give her feathered friends a hug. Glancing around, we find several pairs of longspurs and snow buntings. Near the bank of the slough we frighten a pair of well-camouflaged willow ptarmigan. They lift off across the slough and land on a stack of drilling mud sacks on the opposite shore. A little farther on we spot a pair of tundra swans, with oil wells framing their elegant necks and faces. Birdwatching in the Prudhoe Bay oil fields is just not the same as birdwatching in natural, unadulterated habitat.

I think of snorkeling in Hawaii near a harbor full of sunken garbage and oil slicks contrasted with swimming offshore through a wild coral reef. The same fish may be in both places, but the two diving experiences are entirely different in quality.

We do need harbors, and we do need oil; at least, it seems so in this century. The question is how many, how much, and where? If we have a coastline full of harbors, with no open stretches of beaches, that's poor planning. If America's only arctic coastal plain is designated, in total, for present and future oil development, with virtually no wilderness set aside, that's poor planning. Somewhere we have to draw the line. In the case of Alaska's North Slope, that line should be the western boundary of the Arctic National Wildlife Refuge.

Back in the kitchen trailer, Robin and I prepare a Fourth of July meal for the gang. The gourmet dinner consists of whatever I can pick off the shelves. I pick through the relics and find some Bisquick. One can always make something out of Bisquick.

Robin and I decide to make a pastry recipe shown on the box. It turns out to be your basic glob of dough with a spoonful of blackberry jam in the center. Robin has great fun stirring the dough, spilling it on the table and floor, then poking two or three holes in each pastry. Once the jam is in the pastry centers, she tries to finger it out when I'm not looking. How do you scold a sticky two-year-old when she smiles with blackberry jam on her nose?

We wash a pile of pots, pans, and Texas Ware plates. The Dallas-made dishes must have come with the trailer. As we finish, the wind starts blowing harder than ever. The trailer rocks back and forth, sheets of plywood go flying outside, and the surf's up on the slough. One big gust jolts the trailer, sending spices off the shelves. For a moment the trailer feels as though it might flip over. I worry about whether Dennis will be able to land his plane. A Super Cub can take off at a speed of thirty-seven miles per hour. These gusting winds must be stronger than that. Even if he can land, how will he stay on the ground?

Soon we hear the putter of the Cub's engine overhead. I bundle up Robin, and we run outside toward the lee side of the old warehouse. There's no sign of the helicopter. Dennis circles the camp and makes his approach toward a 600-foot strip of gravel wedged between a long row of trailers and the warehouse. On the approach end of the strip is the slough; on the other end is an oil storage tank with several fifty-foot light poles scattered around. With all the obstacles, it's a difficult place to land in winds gusting above forty miles per hour.

I'm scared. The lightweight canvas-covered Cub could easily blow over once it's on the ground. Why is he landing with only me here to help hold down the plane?

"Don't move," I say to Robin in a stern voice. Sensing my worry, she leans against the building and doesn't budge.

Dennis crosses the slough and approaches the strip at a snail's pace into the wall of wind. I anticipate where he might touch down and run to that point, hoping to grab and steady the wing struts once he's on the ground. In the severe gusts the almost-motionless Cub looks like a seagull teetering above the water in a stiff sea breeze. Within a few seconds Dennis is almost on the ground, then he throttles the plane, pulls up steadily, and circles the camp again.

He makes three approaches, with me huffing and puffing along the strip, ready to grab the struts. I think of those airplane getaway movies, with stuntmen dangling on the wings after takeoff. By the third approach I'm madly waving my arms and shaking my head, trying to signal Dennis that he shouldn't land. (Our air-to-ground radio unfortunately is not working.) Then I hear the welcome sound of the helicopter. Within minutes the chopper lands near the strip, and three bodies pour out the door, sprinting toward me as though in combat. Robin jumps up and down near the building. Ray and Walt take their positions on one side of the strip. Chuck, another technician, joins me on the opposite side. He tells me that they've been in radio contact with Dennis and that Dennis was planning all along not to land until the gang was on the ground to catch the Cub. All of the approaches had been practice runs!

Finally, Dennis touches down, and the four of us are there to make the Fourth of July catch. We steady the plane as Dennis taxis toward the shelter of the warehouse. Robin watches us bring the plane to the tie-downs. She smiles, claps, and shouts "Yea!" Dennis emerges from the plane exhausted, and relieved to be on the ground.

In the morning I awaken from an unusual, vivid dream. I dreamed that I had been thrown somewhere into the twenty-first century, and only a scant amount of wildlands remained on earth. Wilderness had become a prized commodity recognized by all mankind. In the dream I visit the New York Stock Exchange, where wilderness issues have been placed in the neon lights beside industrial and transportation issues. Dollar values have been placed on mammals, birds, and plants, the prices determined by the rarity of species.

Lands that are unblemished from fallout, chemicals, and other forms of pollution are the most highly valued. "Arctic" stock is selling at the highest price per share because the Far North is one of the cleanest, purest zones remaining on the globe. There is so little silence left on Earth that stock buyers are investing in what few noiseless areas remain. Brokers are literally screaming for shares of silence on the floor of the exchange.

In the dream I walk into a room where an old friend is reading a book entitled *The Accounting Principles of Wilderness*. He is taking a business class at some university and is reading a chapter that discusses theories on measuring wilderness in terms of dollars and cents.

The dream is so vivid that I wake up wondering whether there could be any truth in it. Could wilderness become a commodity rare enough that Wall Street would place a price tag on it?

Today we prepare to head for Fairbanks. Dennis leaves early to do some caribou-tracking work before making the four-hour flight back to Fairbanks. Chuck offers to drive Robin and me to Deadhorse so we can catch our jet flight home.

Driving through the Prudhoe Bay oil fields is somewhat like entering a wealthy, private residential area through a security gate, even though the fields are located on public land. The 430-mile web of oil field roads, maintained by the industry, is seemingly closed to public traffic. The industry does grant access to local residents, who live in the small village of Nuiqsut, located just west of the oil fields on the Colville River. They and all commercial operators and industry employees are asked to provide proper identification to security guards located at two checkpoint stations.

At the checkpoints drivers and passengers are asked to give the guards their names, authorized badge or driver's license numbers, and their affiliations. Some drivers must also disclose the property contained in their vehicles. All these statistics are punched into computers. The industry maintains that such high security is needed for safety and liability reasons.

What is most striking about driving through the oil fields is the extensiveness of the development. A 1972 Environmental Impact Statement predicted that the Prudhoe Bay oil field development would encompass a maximum of 550 square miles. Today's fields extend over approximately 800 square miles, and there appears to be no end in sight as far as future expansion is concerned. The Prudhoe Bay oil field is the largest of five reservoirs that have been tapped for production. Other large and smaller reservoirs include Kuparuk, Lisburne, Milne Point, and Endicott.

In addition to these reservoirs is the West Sak field, another Prudhoe Bay–size giant estimated to hold 15 to 20 billion barrels of oil. ARCO has yet to put West Sak into production, and the Ugnu Sands field that overlaps West Sak has not been tapped because the price of oil is too low to make extraction of the heavier-than-usual crude economically feasible. It is anticipated that the industry will develop other promising areas, such as the Colville delta, Gwydyr Bay, Point Thomson, and nearby offshore areas such as the Niakuk region. The state of Alaska has leased about four million acres of the North Slope to the oil industry, and one million acres of offshore tracts in the Beaufort Sea.

Over the next few years those state figures will likely double. Another four million acres of onshore lands to the west, east, and south of the Prudhoe Bay area are scheduled to be offered for lease. Another one million acres of state offshore acreage will be made available in the Beaufort Sea. The federal government additionally plans to lease millions of acres of submerged outer continental shelf lands. To the west of the Kuparuk oil field, beyond the Colville River, lies the 23-million-acre NPRA. This reserve is known to contain approximately six billion barrels of oil, yet industry asserts that it is not economically recoverable at current oil prices.

Even with all these areas available for exploration and production now or in the future, industry argues that the precious 1.5-million-acre coastal plain of the Arctic Refuge should also be opened because it's America's best hope for a major discovery, and that, furthermore, production and export of crude from the refuge would lower the national trade deficit and reduce our dependency on imported oil. Judging from industry's arguments, you'd think America was soon to be down to its last drop of oil and that our national security would be seriously weakened without developing our northernmost wildlife sanctuary.

Pebbles fly into our windshield as we are momentarily enveloped by a cloud of dust from a semitrailer. We emerge from the cloud to see more of the scattered drilling pads, flow stations, work camps, and the distant metallic fences of pipelines and power lines stretching out across the flat pond-pocked tundra. In the distance a skyline of buildings rises above the coastal plain like a city looming above a southwestern desert plain.

"What's that coming up?" I ask Chuck.

"Those are BP's main facilities," he says.

Although the buildings are only a few stories high, they look like skyscrapers from the distance because of the pancake relief. We soon pass BP's operational facilities and its centralized electric plant, which runs on the immediate supply of natural gas. ARCO has similar operational facilities on the eastern edge of the Prudhoe Bay field and within the Kuparuk oil field, about thirty miles west of Deadhorse. North Slope oil companies have one of the world's largest concentrations of turbines, and the industry produces polluting nitrogen oxides in an amount equal to that of a large city. On a day like today, strong winds readily disperse the orange haze, and it blows somewhere else in the atmosphere, contributing to the depletion of the ozone layer.

It is noted in the publication *Oil in the Arctic* that nitrogen oxide concentrations around Prudhoe Bay have experienced a three- to five-fold increase, and sulfur dioxide levels have made a ten-fold jump over a six-year period. Because summer coastal fog and winter ice conditions are frequent in the region, there is concern over the effects of these acidifying pollutants on tundra vegetation, particularly

Aerial of Deadhorse, North Slope oil fields

on lichens. Lichens, an important food source for caribou, are known to be especially sensitive to such pollutants; yet the long-term effects of such air pollutants on lichens and other arctic vegetation are not known.

In a few more minutes Deadhorse comes into view. Although one thousand to two thousand workers are based at Deadhorse, the place looks fairly deserted along the windswept, dusty roads. The settlement consists of thirty-five to forty service companies, two motels, several air carriers, and a couple of landfills and oily waste pits on the outskirts. We drive down the main drag, which is lined with aircraft hangers, rows of modules and surplus equipment, and the motels. No pedestrians are in sight. Chuck drops us off at the airport to await our flight. The small terminal is full of industry workers, and every available seat is reserved on our 737 flight.

Once the jet is airborne, the most noticeable sound is the popping of beer-can tops. Oil industry employees are prohibited from drinking alcohol while based in the oil fields, so the party's on when the workers are homebound. Every passenger near our row has ordered two beers. One fellow tells me that today is shift day for the "grunt workers," or laborers. He tells me that most of the passengers on board work a week, then have a week off.

Scanning the area around Deadhorse, I take one last look at the sprawl of buildings, pipelines, drilling pads, and other facilities that hug the banks of the Sagavanirktok River, and spread west, in a spiderweb-like progression, some fifty miles, to near the Colville River. Glancing north, I spot the five-mile causeway that leads to BP's two offshore artificial islands of the Endicott field.

The Endicott causeway is one of two causeways that finger out into the Beaufort Sea from the oil fields. This causeway slices north through the Beaufort Sea from the mouth of the Sagavanirktok River and has reportedly altered thousands of acres of ocean environment. The natural mixing of salt and fresh water at the river's mouth has been significantly changed. The salinity and temperature are different, sedimentation and erosion patterns have been altered, and prime fish habitat near the river's mouth and along miles of shoreline has been degraded. Two breaches have been built in the Endicott causeway for water movement and fish migration, but they have been only minimally effective.

In a 1986 draft report prepared by the U.S. Army Corps of Engineers, it was documented that under prevailing east-wind conditions at least 60 percent of nearshore habitat along approximately fifteen miles of shoreline has been lost because of the causeway. The loss of this feeding habitat adversely affects anadromous fish such as the arctic and least cisco, broad whitefish, and arctic char.

Similar causeways may extend off the coast of the Arctic Refuge if the 1002 area is opened to development. Although little exploratory drilling has occurred, many tracts offshore of the Arctic Refuge have already been leased to the oil industry, which perhaps anticipated that the necessary onshore support facilities will one day be located within the refuge. Both state and federal governments have offered hundreds of Beaufort Sea tracts north of the refuge under a sort of buy-now-drill-later program.

The jet now follows the trans-Alaska pipeline, a silver ribbon streaming across the tundra, flanked by a gravel highway commonly known as the "Haul Road." Semitrailers, hauling payloads of pipe, lumber, and other equipment, sporadically send up their plumes of dust; they are bound for the oil fields. At a distance the trucks are reminiscent of covered wagons inching their way across a prairie.

I reflect upon my week in Prudhoe Bay. The magnitude of oil development greatly surpassed what I had anticipated. The number of drilling pads and wells, feet of pipe, clusters of facilities, and the amount of garbage and junk were staggering. I did not see any of the thousands of reported oil spills (777 spills, amounting to more than 153,000 gallons in 1988 alone) or the illegal disposal of hazardous wastes. I have only read about such problems. But I did see enough of that far-reaching development to know there is no wilderness there. Enough to know that such development has no place now or in the future within the Arctic National Wildlife Refuge.

❖

Update: January 2011. Arctic oil development has increased substantially on North Slope state lands since 1990. In 1989 there were six producing fields on the North Slope. Today there are thirty-four producing fields whose network of roads, pipelines, drilling pads, and facilities encompass more than one thousand square miles. As of November 2010, the Alaska Oil and Gas Conservation Commission has a record of 2,444 oil production wells on the North Slope. There are also many discovered oil and gas fields that have not yet been developed.

In October 2010, the state of Alaska leased 725 North Slope and Beaufort Sea tracts to the oil industry, amounting to 2.7 million acres—a bigger area than the entire Arctic Refuge coastal plain. All of these tracts have moderate to high potential for discoveries. This recent sale brought in $63 million in leasing bids to the state treasury.

Over the next five years the state of Alaska plans to conduct annual North Slope and Beaufort Sea area-wide lease sales, offering more than 14 million acres of moderate to high potential tracts. These include sensitive offshore areas within three miles of the arctic coast that are under state jurisdiction. The Arctic Refuge coastal plain remains the only small sliver of federal coastline that is off-limits to industry.

Although industry attempts to avoid accidents, there are an average of 450 oil and toxic spills each year on the North Slope. Drilling contractors have illegally dumped hazardous wastes, facing serous criminal penalties and fines. In 2006 an aging, corroded pipeline leaked 212,000 gallons of crude onto the tundra, the largest North Slope spill and due to BP's negligence. In 2009 another overpressured pipeline exploded, spilling 45,000 gallons of crude and water at BP's Lisburne Field. Oil spills are the nature of the beast.

11

Coming Home

WE STEP OFF THE TWIN-ENGINE Navajo at Barter Island into surprisingly balmy air. The airstrip sits on a finger of land that juts into the Beaufort Sea and points toward the North Pole. Sea ice gnaws at the shoreline within a few yards of the plane, yet the temperature is 45 degrees. This is T-shirt weather in the Arctic.

On past summer trips to the Arctic National Wildlife Refuge, we jumped off the plane into the bite of an arctic gale, with temperatures in the teens. I'm thankful that it's calm and relatively warm, for this is our daughter's first trip to the refuge.

After a long day of transition, it's exhilarating to stand on the edge of North America once again. I look beyond the gravel airstrip which is lined with rusty oil drums. The white sea of pack ice shimmers endlessly to the north. Gazing across the ice, I draw an imaginary line between us and the North Pole, understanding that no human stands between the two vantage points, unless some polar expedition is in progress. Our closest human neighbors beyond the pole are the Laplanders, almost three thousand miles away.

Beyond the small Inupiat village of Kaktovik and the DEW line site, the coastal plain of the Arctic National Wildlife Refuge sweeps south, free of human obstacles, to the glaciated peaks of the Brooks Range, about forty miles away. The only sign of civilization across this giant sweep of mountain, plain, and sea is the lone dot of Kaktovik and the DEW station, with its two microwave towers rising above the plain like a pair of giant Mickey Mouse–shaped ears.

Don Ross, a friend and pilot, meets us at the airstrip. Earlier, he had offered to fly Robin and me into the Aichilik River drainage. Dennis, under contract with the USFWS, is busy flying caribou surveys. He is somewhere over the coastal plain of the refuge tracking members of the Porcupine caribou herd from his Super Cub. Robin and I plan to camp alone on the Aichilik River for a couple of weeks, but we hope that Dennis will stop in to see us whenever he can.

Don's Cessna 185 is packed to the brim with camera gear, backpacks, diapers, and containers of food. Japanese photographer Michio Hoshino, who is working on a caribou book, is also with us. Michio sits in the only available front seat. Robin and I wedge ourselves into the back of the plane. There is no room for standard aircraft seats. We sit on Michio's backpack, attempting not to squash his fruit, which is stashed in one of the pockets.

It feels good to get off the ground, knowing that our next stop will be the Aichilik River, about fifty miles away. We fly over Kaktovik's small rectangular patch of buildings fringing the Beaufort Sea. Things have not changed much since my first visit to Kaktovik eleven years ago. There are a few new brightly painted frame houses on the west side of the village, and a new high school and fire station, both built several years ago with funds generated from oil revenues. The same cluster of driftwood-gray plywood boxes is in the old part of town. Debris of all kinds is scattered in between, everything from whale carcass remains to discarded snow machines, scrap wood, and rusted oil drums. There are no lawns, just a few patches of tundra. The plane banks away from the tiny settlement hugging the white sea.

The people of Kaktovik have mixed feelings about oil development out their back door. While some would welcome the boost in the local wage economy, others feel that the magnitude of such a project would bring a host of social problems to the village, particularly with respect to the increased use of alcohol and drugs. Many villagers are concerned that their access to subsistence hunting areas would be restricted by a major oil field. Such is the case in the Prudhoe Bay oil fields, where the area is closed to big-game-hunting, and the access highway is closed to hunting with firearms within five miles of the road. Other villagers are worried that such widespread industrialization would pollute their homeland. In the words of Jane Thompson of Kaktovik during a recent USFWS hearing:

> . . . we're not going to be able to travel freely like we do now. We're going to be restricted. . . . I value our land up there so much. It's where we go hunting and where we get our food. There's clean water all over. It's pure land. Once it's open, it's just not going to be the same. We're going to be treated the same way like [in] Prudhoe Bay.

As we cross the expansive coastal plain the engine's drone quickly puts Robin to sleep. The open, treeless plain averages about twenty-five miles in width, pinched between the sea's shorefast ice and the Brooks Range rising to the south. It is a relatively narrow band of wetlands and rolling tussock slopes, particularly when compared to the pond-pocked coastal plain zone south of the Prudhoe Bay oil fields one hundred miles to the west. From Prudhoe Bay, the coastal plain stretches roughly one hundred miles inland before reaching the foothills that skirt the Brooks Range. When standing on the ground at Prudhoe Bay, you cannot see the mountains. In contrast you can see the Brooks Range from any vantage point on the coastal plain of the refuge, weather permitting.

On this clear day, viewing the Brooks Range from the coastline is similar to what it must have been like to view the southern Rockies from the Great Plains before there was a Denver. The Brooks Range rises dramatically from the plain, and we can see the faint sparkle of glaciers cradled among the east–west rampart of lofty peaks. With shimmering tundra ponds racing below us, we have the unique opportunity to gaze across a wildlife-rich arctic prairie, while winging our way from the edge of the North American continent toward the northernmost extension of the Rocky Mountains. Where the Great Plains once teemed with countless buffalo, this arctic prairie offers an ancient sanctuary for what can be considered the greatest wildlife spectacle remaining in America, the aggregation and migration of the Porcupine caribou herd.

Summer is just arriving on the coastal plain. The tundra grasses and sedges are beginning to show their new emerald green growth through winter's brown cloak. We fly over a high-density area of polygon-shaped ponds framed by spongy bands of tundra. The zone reminds me of rice paddies in Southeast Asia. Many nesting shorebirds are sprinkled across the miles of wetlands, but we're too high above the ground to identify the specks, although we do spot a few pairs of the large ivory-white tundra swans. Where the continental United States has lost at least half of its productive wetlands habitat to industrial development, and continues to lose hundreds of thousands of acres each year, it's comforting to know that this northernmost zone remains untouched, yet the threat of oil and gas development here is ever-present.

Within a few minutes of departure, we see our first group of several hundred caribou, mostly cows with their month-old calves. The bull caribou are striking in their dark coats of new fur, while the cows still wear the bleached-white remnants of their winter coats. Within a few weeks the tattered-looking cows will begin to grow their new coats and antlers. It is believed that the cows delay their fur and antler growth because they put such tremendous physical and hormonal energy into calving and lactation during the early summer period, with little strength left for hair and bone production.

As we move inland, the tussock-covered tundra begins to dominate the coastal plain. From the plane the tussocks look like goose bumps on weathered skin. The closer we come to the Aichilik River, the more I feel like I'm coming home. For many years Dennis and I have been lucky enough to spend a large portion of our lives living, working, and recreating within this refuge. Now, for the first time, I have a special opportunity to share this legacy of wilderness with our first child, without flying over oil rigs to reach our destination.

Just before reaching the Aichilik River, we pass over a small group of about twenty muskoxen; they don't seem to be disturbed by the roar of the engine. Don thinks they must be getting accustomed to the aircraft connected with the numerous scientific studies of the coastal plain and with the oil and gas assessment work. These shaggy survivors of the Ice Age indeed look like "walking haystacks," as Ave Thayer would say.

We soon approach a small tundra strip on the west bank of the Aichilik River about four miles north of where the coastal plain begins its roll to the sea. We've just entered the 8-million-acre wilderness-designated portion of the refuge, although we still have a clear view of the coastal plain's 1002 zone that may be opened to full oil and gas exploration and development if Congress so deems. During this very week congressional hearings are being held on the results of the Department of Interior's 1002 report, which recommends that the area be opened to full-scale oil exploration and development.

Dust and gravel fly for a few moments as we touch down and wheel to a halt. I'm anxious to unstrap and jump out. Our home for the next two weeks lies in the foothills of the Brooks Range, in a wildlife-rich transition zone between the mountains and the plain. From our location on the Aichilik, we can look north beyond the opening of the valley, across the coastal plain, and south to the higher peaks along the spine of the Brooks Range. Broad, gently sloping valleys flow toward us from the east and west. Our vantage point offers spectacular views in every direction, and the site is ideal for observing wildlife moving up and down the valley.

The tundra strip lies on the old floor—which has not been flooded for many years—of the Aichilik River. Near the strip is a noticeable overgrown cutbank, about twelve feet high and brightly covered with blooming white heather, violet shooting stars, yellow cinquefoil, and bearberry bushes. Along this former route of the river, the tundra is exploding with brief summer color.

Robin is delighted to wake up when we emerge from the plane. She stands on the tundra and bends her knees a few times, testing the ground's spongy texture. Then she squats down and looks at the new tapestry of plant life. The dwarfed arctic plants are built to her scale. When I see her smile at the tundra, her face within inches of a white dryas flower, my worries of bringing a toddler to the Arctic dissipate.

Within a few minutes the plane roars off, quickly becoming a puttering speck in the distance. Don will make a couple of trips shuttling Michio and a second photographer to another river valley. Then Robin and I will be truly alone, with miles upon miles of wilderness surrounding us.

Within an hour a small group of caribou moves into the Aichilik River valley from the west, down the naked tundra slopes that grace the Egaksrak River. They trot along the opposite slope from our camp, heading toward the coastal plain, where most of the Porcupine caribou herd is located. Many of the dozen animals filing by us are cows and calves, although there are a couple of bulls mixed with the group. From my arms, Robin silently watches her first caribou.

As we set up camp, Robin is excited about her new surroundings. She bellies along the tundra, touching, smelling, and tasting all the new Arctic plants. Like any toddler, she equally uses all her senses when discovering a new world. Bearberry leaves, lichens, mosses, dwarfed willows, blooming dryas, and last summer's withered leaves are densely matted together, forming an intricate puzzle of surprises for her. After setting up the tent, I study the surroundings of our new home more closely. From the floor of the Aichilik valley, we are shouldered by gray, rounded ridges that gently rise from the tundra. Smooth-lined slopes characterize a very weathered range of foothills that gracefully front the Brooks Range. These northern ridges, once scoured by glaciers, have been exposed to some of the most extreme climatic conditions in the world. Centuries ago this region carried the weight of an ocean. Over the millennia, these ridges and mountains have been lifted, tilted, and thrusted into the rawness of a polar world.

Green rivulets of alpine tundra spill down one ridge, like topping on a sundae. The falling ribbons of tundra intersect with numerous caribou trails that zigzag across the ridge's face. These animal etchings look like nature's road map turned upright.

Occasional outcroppings composed of more defiant rock erupt from some of the nearby surrounding ridges and mountains. Like lonely glacial erratics, they break the smooth scan of the eye. A golden eagle soars over a distant outcropping in search of its prey. Scanning the floor of the Aichilik's green valley, I see several weathered gray hills in the distance rising from the deceptively smooth tundra the way the backs of gray whales rise above a calm sea.

Once again we've returned to the land without sunsets; there are only sunrolls. As evening approaches, Robin and I watch the golden orb roll behind a ridge for a time, then reappear low along the northern horizon, casting an amber glow on the tundra. Robin seems puzzled by her shadow, which has grown long in the midnight sun.

When we crawl into our sleeping bags, it feels good to be on the ground once again, positioning our bodies around the tundra's bumps, snug in our dome tent. We are truly alone in this most remote northern wilderness, with only a layer of nylon separating us from the wild.

A southeast breeze caresses our tent as I drift off to sleep. The soft steady churning of the Aichilik River blends with the music of golden plovers, Lapland longspurs, sandpipers, and redpolls. With no nightfall we will hear bird songs throughout the twenty-four-hour day.

The wind's gentle nature seems to lift yesterday's worries and trivia, blowing them into a distant, less urgent perspective. It is easy to become trapped in the fast pace of urban life and to be troubled by the never-ending list of things to do, however mundane those things may be. The rush-hour traffic, long lines at the grocery store, or unfinished chores can distract and drain us so that we lose awareness of the more important aspects of life. We can become so busy that we forget to appreciate, or even observe, the beauty of the sunrise.

The wind cleanses the mind by blowing away layers of dust that smother the brain's senses, dust that has settled over time from the grinding of civilization. Brain dust smothers our acuity and reduces our capacity to hear silence and the wind's gentle song. Our deadened senses, accustomed to the hum of the refrigerator, the whir of the fan, and the ticking of the clock, need revitalization. In this arctic wilderness our senses are sharpened. The drone of civilization is left behind. The wind opens windows to the mind that have been closed for too long.

Through the night I am periodically awakened by a loud thumping sound. At first I picture a grizzly bear outside the tent ripping up huge chunks of tundra and tossing them like bales of hay. I had noticed a few bear diggings around our campsite. The thought of a foraging bear, cratering the tundra outside the tent door sends my heart racing.

After the third loud thump, resembling the sound of a deep bass drum, I realize it is only sections of ice calving into the river. This is a small sheet of two-foot-thick *aufeis* located near camp. With the warm temperature, seventy to eighty degrees, the ice is melting rapidly.

On July 2, we awaken to a clear, beautiful day, with a gentle breeze to keep the mosquitoes grounded. In the early morning light the soft white plumes of cotton grass in the surrounding valleys could be mistaken for fields of daisies. Cotton grass thrives in this tussock-covered valley. In the distance the valley looks as if it is freshly dusted with snow, but the "snow" is really miles of backlit cotton grass.

While we eat breakfast, a group of twenty bull caribou and a lone cow file down the river. Robin has a close look at all of them. She is wide-eyed and occasionally points at the animals while whispering, "Ahhhh . . . Ba." The group is en route to the coastal plain, walking at a steady gait, passing two strangers without noticing.

The bulls carry new sets of velvet-covered antlers which appear to waver slightly as they walk. The nonsolidified bone, dense with blood capillaries, has not fully hardened. The ornate antlers will lose their sheaths of cattail-like fur by the end of August and gain their protective calcified bone in preparation for the fall rut.

Robin and I take a morning hike up the valley along the river's edge. I do most of the walking, while Robin has a free ride on my back. We pass several semipalmated plovers as they scurry along the gravel bars near their camouflaged ground nests. Plovers remind me of Charlie Chaplin in fast motion, with legs the length of toothpicks. They sprint a few yards, then freeze for a few moments, as if struck by lightning, then race on again.

Semipalmated plovers received their name for the simple reason that they dash around on half a palm. Their outer and middle toes are partially webbed to the second joint. This small shorebird, with the distinguished black ring draped around its neck, migrates as far south as Patagonia for the winter months.

At the moment the plovers appear to be Robin's favorite bird, and she tries to mimic their piping. "Peep . . . peep," she calls to them, reaching toward them from the pack. In addition to the plovers, the gravel bars and willow bushes are full of other birds, such as the snow buntings, several species of sandpipers, and redpolls, who constantly chatter as they flit from bush to bush.

On the way back to camp, we spot a cow caribou heading up the valley. I wonder if she is in search of her missing calf. Later, back at camp, we spot a lone caribou calf on the opposite side of the river, searching for its mother. Its nose to the ground, the calf trots up the valley, stopping here and there to look around; then it urgently moves on.

In the evening we crawl into the tent as a cold breeze begins to funnel up the valley from the north. We fall asleep in our woolen hats with our sleeping bags snug around our chilled faces. I sense that we're in for a change of weather.

For the past two days we've experienced a steady arctic gale out of the northeast. Seldom has it let up. Robin takes a few steps outside the tent, then gets blown off her feet onto the tundra. She quickly learns that we need the tent's shelter. We spend many hours reading six children's books, again and again, until I can recite them from memory. When I begin to go crazy from the repetitious reading, we play every possible game that can be invented for a toddler in a tent: hide-and-seek in the sleeping bags; gymnastics, with me as a balance beam; and tent basketball—throwing a tennis ball into a cooking pot, boot, or hat.

The sky is crystal clear, yet the oppressive wind keeps us tentbound hour after hour. There must be a high pressure system centered offshore somewhere above the frozen Arctic Ocean. When we poke our heads out of the tent, we can feel the pack ice on the teeth of the wind.

A good arctic blow reminds me why tundra plant life is so dwarfed. Any plant daring to grow more than a few inches above the ground falls prey to the wind. The mountain avens, for example, is well designed for the strongest of blows. Its short stem is flexible, bending to the ground the way our fiberglass tent poles do. From the tent Robin and I watch hundreds of these dainty white flowers hammer at the tundra, their petals shivering against the ground.

In the blow, willow bushes crawl across the tundra around our tent. They look like old hunched-over men, struggling across the land on a long pilgrimage. The white-bearded backs of the willow leaves reflect the midday sun the way window frost reflects early-morning light.

After two long days, the wind finally begins to calm in the evening. I walk up to the tundra bench above our camp while Robin sleeps. The brutal force of the wind has damaged some of the taller plants. Arctic poppies, whose yellow heads follow the arc of the sun, took the worst beating. Their long graceful stems have been snapped by the wind about four inches above the tundra. Like broken masts at sea, the stems have been almost sheared off by the gale. Withering yellow flower cups dangle upside down on their splintered vines.

Later my adrenaline level rises as midnight sunlight pours down the valley. The Aichilik valley of a few hours ago is transformed: Colors that were washed out in the midday sun are richer, more vivid. Faded greens are deepening to the color of a spring-green meadow. Yellows are turning to a brilliant gold. Dull gray mountains take on new relief and a gradient of colors ranging from soft lavenders to charcoal blacks. Lengthening shadows reveal outcroppings, ridgelines, and distant valleys that were invisible in the bright sunshine. The midnight light brings the smallest plants and the highest peaks into three-dimensional viewing. The tundra is no longer flat, the mountains no longer sheer. The light transcends detail the way a good topographic map does. To sleep during the midnight sun is to miss the Arctic at its fullest moment of splendor.

I walk along a gravel bar that only a few hours ago was largely a drab gray. Rounded river stones had looked uniform in color and dimension. In the late evening light the same bar reveals an entirely different set of rocks and varied colors: Navajo reds; jade greens; tawny browns, with flashes of white sedimentary strata; and some lavender hues. The gravel bar is alive with eons of geologic stories.

I pick up a fossilized coral rock and once again marvel at the thought of standing on an old ocean floor. Another rock has wavy sedimentary strata, layered over

Hiking with Casey

time by water lapping upon sand deposits. The vastness of the surrounding arctic landscape makes me feel like an insignificant speck of human life, and these rocks place humans entirely off the geologic time chart.

Along a sandbar near the edge of the river, a drift line of white caribou hair has formed where the waters have receded. It reminds me of eelgrass left by an ocean tide. Caribou tracks cover every inch of the bar. I had heard that a group of forty thousand caribou crossed the river not long before we arrived. I wonder if any of the calves drowned during that particular crossing, or how many may have become separated from their mothers.

To the north the sun rolls along the horizon above the coastal plain. Squinting in the low-angle light, I gaze down the valley to where the last foothills drop to the open plain. The Aichilik River catches the sun's glare, sending a shimmering ribbon of liquid silver toward the Beaufort Sea. I watch the river curve from bank to bank, until it becomes as narrow as a pen's stroke near the mouth of the valley. Two pintail ducks fly by me, a few feet above the water, in the midnight glint. They ride upon a thin cushion of air, following the bends of the river, soon becoming fluttering silhouettes in the sun.

We awaken early from intense heat, like two pots baked in a kiln. With temperatures in the 80s, this is the hottest day so far. It is dead calm, and I suspect the caribou must be running for the coast to cool off and seek relief from the swarming insects. This buggy-hot day should drive the caribou together to begin their post-calving aggregation. I suspect Dennis is out flying with other biologists, closely tracking the herd's movements in preparation for a census count.

Robin and I walk up to a bluff above camp to scout for caribou. The mosquitoes form thick, aggravating clouds around our heads, and without bug repellent we would be devoured. With little precipitation in recent weeks the tundra is already dry, and many plants are going to seed. The grasses and sedges are so crisp that they crunch beneath my feet; it's as though I'm walking on soda crackers.

During our first week on the Aichilik, we have watched clumps of violet shooting stars reach their blooming peak, wither, and die. The white heather along a southeast-facing cutbank has come and gone. The tiny bell-shaped blossoms have turned to miniature paper lanterns.

We discover a caribou carcass from last winter. The rib cage is partially intact, and its two jawbones lie face-to-face. Caribou fur is scattered everywhere, and wolf tracks surround the site. I wonder if the animal was killed by wolves or by another predator, or if it died of natural causes.

Back at camp we drink lemonade by the quart and create some shade by placing Robin's poncho over the spotting scope tripod. While we sweat in the heat, four ravens fly just over our heads and cackle at us. Robin looks up smiling and waves at them as they make cartwheels in the sky. She calls to them, "Da . . . dai, da . . . dai." It is the first time Robin has ever observed ravens at close range; and the first time she has ever spontaneously waved at any creature, including man.

After Robin's morning nap, it is hotter yet. We walk to a sandbar along the river, and Robin makes prints with her tiny feet next to the caribou tracks. A quick plunge in the frigid water cools us off. Robin is delighted that she can run naked and freely on the sand, looking at animal tracks and bird feathers, picking up rocks.

In the late afternoon we are still drinking lemonade and watching cumulus clouds build up over the mountains. These are the first clouds we have seen in more than five days. I hope they will come our way and provide some needed shade.

Robin and I sit in what little shade the tent and hanging poncho provide. Even though she is well covered with baby sunscreen, I still worry about the intense sun. Suddenly I notice something moving in the willow bushes. The hump of some blond, furry animal is about three hundred yards from our tent. For a moment I think it's a small grizzly, and my heart starts pounding. Within a few seconds I see it's a wolf with its head lowered, pawing at the ground. At first glance, all that was visible was its light-colored back.

The wolf gradually walks our way; we remain motionless next to the tent. Robin spots the approaching creature, a dog in her mind, when it is about one hundred yards from the tent. The wolf draws closer and closer, walking very slowly with its tongue hanging out, panting in the midafternoon heat. Its fur is bleached white and mangy as it is shedding its old coat. The wolf looks tattered and tired after a long day on the prowl.

The closer the wolf approaches, the more excited Robin becomes. She stands on her feet and starts to babble. I explain to her that it's a wild dog and tell her she can't pet it. That makes no sense to her. She wants to charge over to the animal and give it a big hug, so I grab the back of her T-shirt to hold her in place, and caution her in a whisper to keep still.

The wolf approaches within forty yards of the tent (I paced it off later). This is the closest encounter I've ever had with a wolf, and I'm mesmerized. As the wolf comes parallel with us, Robin calls out in a loud voice, "Da . . . da . . . da!" and reaches out toward the wild dog. I think she expects the wolf to walk over to her for a pat. She calls a second time, "Da . . . da," and the wolf stops and stares at Robin for a few moments, although it feels like hours.

In all our years in Alaska, I have heard only one account of a wolf attacking a trapper, yet I still worried that Robin might appear to be just the right-size prey, and somewhere in the back of my mind I remember reading about a wolf species in India that reportedly nabs little children. Fact or fiction? I guess with some certainty that this wolf has probably never seen a human toddler and is probably just curious.

After a long stare at Robin, the wolf continues on past camp, walking a weary pace. A few moments later Robin calls a third time, and the wolf turns around and looks at her again briefly, then proceeds up the valley. Robin quietly watches the wolf until it is completely out of sight.

This particular wolf is a member of one of at least six known packs that reside in the refuge on the north side of the Brooks Range. It is roughly estimated by ADF&G that 5,200 to 6,500 wolves reside in Alaska as of 1988, and these numbers fluctuate, depending on the availability of prey and on annual mortality. Anywhere from two hundred to three hundred packs of wolves may roam throughout northern Alaska, and their territories are extremely variable, depending on food sources. If a wolf pack lives in a valley where it has easy access to a residential moose or sheep population, the wolves may have a very small home range. Other packs may have to travel great distances to survive, following migratory caribou or preying on whatever small game they can catch.

Although Canada, Alaska, and the Soviet Union have largely stable or increasing gray wolf populations, the status of gray wolves and other subspecies throughout most of the continental United States and around the world is very poor. The historic range of gray wolves in North America once extended throughout the vast majority of America's states, as far south as northeastern Florida and southern Texas, and into central Mexico. Over the years gray wolves have been eliminated by humans or pushed out of their former range because of agricultural and industrial development.

Minnesota is the only state of the lower forty-eight that has a well-established gray wolf population, approximately twelve hundred to fifteen hundred wolves. Wisconsin has about twenty-five wolves that are part of a reintroduction program, and about fifteen wolves have moved south from Canada into Glacier National Park in Montana, reestablishing a small population. Efforts may continue to reintroduce wolves into other areas of their historic range, although there is much controversy over the issue, particularly among livestock growers.

Wolves once extensively occupied most regions in both the Old and New Worlds, and as in the United States, they have been eliminated from most of their former ranges on the planet. At a 1988 international wolf symposium, it was reported that many wolf populations around the world are either threatened or in danger of extinction. Norway and Sweden share a total wolf population of 11 animals, while northern Finland has 10 to 20 wolves, with another 60 who live along the Soviet Union–Finland border. A few hundred wolves live in Italy and Israel and feed primarily at garbage dumps. Portugal has about 150 to 200 wolves, but that population is declining because of decreased habitat. Spain and Poland have relatively stable wolf populations; each country has about 800 to 1,000 wolves. An estimated maximum of 50 wolves remain in Mexico, where habitat loss and competition with the cattle industry have reduced their numbers. Wolves continue to be trapped, poached, and poisoned throughout the world, and their future can be considered bleak.

Although millions of square miles of habitat are available for an estimated thirty to sixty thousand gray wolves in Canada and for those in Alaska, there are still conflicts between humans and wolves. Of the nine provinces in Canada, six provincial governments conduct predator-control programs to protect livestock, and three provinces use predator control for wildlife management. Wolves no longer occupy their former range in the southern portions of Alberta, Saskatchewan, and Manitoba, where much of their habitat has been lost to agricultural development.

Within Alaska, the greatest conflict between wolves and man is hunting competition over moose or caribou. In past years, when moose or caribou populations have diminished as a result of sport or subsistence hunting pressure, wolf and bear kills, or severe winters, the ADF&G has enacted controversial wolf-control programs, largely through aerial shooting, in an effort to elevate prey populations. Such control programs have never been proposed or enacted within the federally controlled Arctic Refuge.

Hunting and trapping of wolves is allowed within the refuge, and illegal poaching does occur. Since Alaska's northeastern corner is so remote and there are few human residents, wolf numbers have remained relatively stable over time. Yet there have been documented reports of illegal aerial shootings that have wiped out entire packs in some drainages. Also, the rabies virus has killed off a number of wolves on the north side of the Brooks Range.

Given the fact that wolves have become threatened, endangered, and have disappeared in many parts of the world, it is a rare experience to be able to watch, and in the case of Robin, talk to, a member of one of the northernmost wolf packs in the United States. The Arctic Refuge, and other northern undeveloped regions, provide the last stronghold for the gray wolf.

"We are lucky people to see that wolf walk by our tent," I say to Robin, who still seems puzzled by the event.

In the late evening I drift off to sleep watching Robin's innocent face in her adult-size sleeping bag. I think about wolves, grizzly and polar bears, Dall sheep, and the many birds that have inhabited this refuge for centuries. I wonder if one hundred years from now this northeastern corner of Alaska will still be a wildlife and wilderness sanctuary, for my grandchildren and great-grandchildren, or will our insatiable appetite for resources and our mushrooming world population swallow this landscape and eliminate its free-roaming residents.

If history is any indication, the future for wilderness areas is not promising. Three hundred years ago 100 percent of America was wilderness. As of 1988, only a small portion of the country remains in a wilderness condition, with 3.8 percent of the nation legally protected under the Wilderness Preservation System. Most of that designated wilderness lies in Alaska. Much of America's landscape has been altered or disposed to agriculture, industry, transportation corridors, and sprawling cities.

Around the world industrialization continues to spread to some of the most remote corners of the world. On behalf of the 1987 Fourth World Wilderness Congress, the Sierra Club conducted a world wilderness inventory and mapped remaining wilderness areas by continent. That study reveals that about one-third of the earth is still wilderness (the study identified wilderness blocks of one million or more acres showing no evidence of development). Of that one-third, 42 percent lies in the high Arctic or in the Antarctic. Antarctica and Greenland are nearly all wilderness. Another 20 percent of the planet's wilderness lies in warm deserts, largely in Africa. The remaining 38 percent of wilderness (or about 13 percent of the earth's total landmass) is found largely in temperate regions and in the tropics, and a small amount in mixed mountain regions.

Where does the Arctic National Wildlife Refuge fit into this global picture? According to this first world inventory of undeveloped lands, the Arctic Refuge is clearly the largest wilderness area in the United States. On a global scale, the Arctic Refuge, combined with the Northern Yukon Park, is one of the largest remaining wilderness regions on Earth. There are wilderness zones within Canada and Siberia that are larger, particularly in the Northwest Territories and in northern Siberia. And, as already noted, most of Greenland is wilderness.

However, the Arctic National Wildlife Refuge and the adjacent Northern Yukon Park stand alone in one significant respect. Of the established conservation units on this planet, the Arctic National Wildlife Refuge and Northern Yukon Park compose the largest protected block of wild habitat in the world. Furthermore, there is no other area that provides sanctuary for such a wide diversity of arctic and subarctic species and for the wide spectrum of habitats contained within the refuge. It has been proposed in the past, initially through the work of George Collins and Lowell Sumner in the 1950s, and later by Canadian and U.S. conservation groups, that this precious region be given full protection by designating the area as an International Wilderness Reserve, or as an International Biosphere to preserve the entire ecosystem, as well as the subsistence lifestyles and cultures of the aboriginal people who have lived in this northern region for thousands of years.

With the imminent threat of oil and gas development in the Arctic Refuge, such an international classification is vital for the preservation of one of the world's greatest natural treasures and wilderness zones. As our wild areas shrink and species continue to be extinguished at the staggering rate of one thousand species per year—as reported by The Nature Conservancy—the time for such a classification is now. Instead of draining America's domestic supply of oil and tearing up the most biologically productive part of this refuge for profit, we should think of our children, wisely conserve our nonrenewable resources, and preserve this rare sanctuary.

Although about 70 percent of our nation's daily supply of oil is consumed by the transportation sector, we are once again building bigger cars with larger engines and raising the speed limit, both of which increase fuel consumption. Information contained in a graph produced by The Wilderness Society indicates that if the U.S. fuel efficiency mileage standard were raised on new car models from 26 miles per gallon to 27.5 miles per gallon, more oil would be saved than the anticipated production within the Arctic Refuge over the projected thirty-year life of the oil field.

The Department of Interior estimates that there is only a 19 percent chance of finding economically recoverable oil within the refuge and that the average estimate of potential oil reserves is 3.2 billion barrels, far less than one percent of worldwide ultimately recoverable oil resources. Over the life of such an oil field, average production would contribute to a mere 1.8 percent of total U.S. consumption. Fighting increased oil imports with Arctic Refuge–produced oil is, in the words of Christopher Flavin of the Worldwatch Institute, "like trying to stop a major fire with a teacup."

Our nation's energy policy appears to be based solely on the momentary price of a barrel of oil, rather than on looking into the future, advancing our current technology relating to alternative sources of energy, and developing more efficient consumer products. While oil is cheap and lines at the pumps are gone, consumers are burning up more fuel, as though the oil is a surplus bumper crop. The only difference is that crops continue to yield fruits and vegetables over time, while our finite supply of oil cannot be replaced. As I watch Robin sleeping, I think of the words of David Brower. Instead of pampering ourselves with conveniences and depriving future generations by unwisely consuming all our oil, "we should stop stealing from children." Oil and wilderness have one thing in common: there is a finite amount of both. What we take out of the ground, we can't put back. What wilderness we alter or destroy, we can't re-create. If we industrialize the wildest corner of America for temporary economic gains, we are robbing from future generations of mankind and wildlife.

I look out through the tent door, beyond the shadowed foothills, across the vast stretch of coastal plain, knowing that the decision of whether to open this northernmost sanctuary to development rests with Congress.

New Afterword

THIRTY-FIVE YEARS AGO, I first set foot in the Arctic National Wildlife Refuge while teaching in the Athabaskan Gwich'in community of Arctic Village. Stunned by the beauty, humbled by the wandering caribou, we were embraced by a magnificent wilderness. That truly wild country inspired me then, as it still does today.

Since 1975, I've hiked, snowshoed, skied, and floated through the Arctic Refuge on many journeys with family and friends. Each trip always brings new discoveries, vistas, wildlife encounters, storms, adventures, and the challenges of camping in a remote wilderness. Be it an arctic storm that confines you to a tent, a dramatic mountaintop vista, a tuft of flowering moss campion, the clicking of ten thousand caribou hooves, or a grizzly bear crashing through the willow bushes, each waking moment is savored.

Good news. The Arctic National Wildlife Refuge is still wild and free.

Yet twenty years have passed since I wrote this book and there have been changes, most notably associated with global warming and increased oil development on state lands to the west of the Arctic Refuge. Dramatic climate change has affected both the landscape and the wildlife of the Arctic Refuge and Beaufort Sea.

The glaciers that we climbed in the Romanzof Mountains during the 1970s have retreated. During the past half-century, the McCall Glacier and other alpine glaciers in the Arctic Refuge have receded dramatically. The summer melting of glacial ice has steadily increased. If ice loss continues to accelerate, all Brooks Range glaciers could disappear within this century.

Indeed, temperatures are warmer. Over the past fifty years winter temperatures in the Arctic, both in Alaska and western Canada, have risen as much as five to seven degrees Fahrenheit. Rain precipitation during the winter months has also increased. Midwinter icing caused by freezing rain has reduced access to food for animals such as the muskox and caribou. This may be a potential factor in the serious decline of some muskox and caribou populations. In 1990, there

Journal writing beneath Mount Michelson

were 350–400 muskoxen in the Arctic Refuge, and today fewer than 50 can be seen in any given year. Some of the muskoxen have dispersed to other regions in the Arctic, but others have perished due to bear predation, disease, or changes in vegetation.

Some caribou herds in Alaska and Canada are experiencing downward trends while others are stable or growing. The Porcupine caribou herd has dropped from an estimated 178,000 animals in 1989 to about 169,000 in 2010. State biologists report that the herd is stable. While populations can rise and fall as part of the natural cycle, some herds are experiencing rapid decline due to climate change or loss of habitat. Canada's endangered Peary caribou of the high Arctic have been in serious decline since the 1960s, largely due to extreme weather events that bring deep snowfall or freezing rain, cutting off their food supply.[7]

Polar bears, including those that begin their lives on the Arctic Refuge's coastal plain, are now a threatened species. They face the perils of global warming head on. Thinning sea ice, longer ice-free seasons, and the loss of sea ice habitat have dramatically affected Beaufort Sea polar bears. Struggling bears have drowned in large expanses of water or become emaciated. Recent studies show that the survival rate of cubs is much lower, and body weights have diminished. As a young girl once said during an arctic rally on Capitol Hill: "Polar bears are good swimmers but they can't swim forever."

In 1990, about one-third of the Beaufort Sea pregnant polar bears chose to den on land. Today, two-thirds of the bears select land over sea ice for denning, due to deteriorating ice conditions. This increasing trend to den on land makes the coastal plain of the Arctic Refuge more important as critical habitat for our threatened bears. Thirty years of continued studies also show that the Arctic Refuge coastal plain has the highest concentration of land-denning Beaufort Sea bears in both Alaska and Canada.

What will happen to the polar bears? Polar Bears International reports that the world population of polar bears is estimated at 20–25,000, and as many as two-thirds of these bears may disappear by 2030 because of the rapid loss of sea ice. Other threats include pollution, poaching, and industrial activities.

In light of the massive Deepwater Horizon oil spill in the Gulf of Mexico, should our federal and state governments consider allowing offshore drilling in arctic waters, where there is no proven technology to clean up an oil spill in sea ice conditions? Should the polar bears, seals, walrus, and other marine life be subjected to this type of high-risk industrial activity, when they are already coping with the dynamics of climate change and the loss of their sea ice world?

We have other energy sources and choices to curb our insatiable appetite for oil. We can be more responsive, more courageous, and more innovative, and leave sensitive places like the Arctic Refuge and the Beaufort Sea wild and free.

One recent study gives us cause for concern and for hope. A group of scientists gravely note that if greenhouse gases continue to rise as projected, all of our polar bears may be extinct by the end of the century because of rising temperatures and loss of sea ice. But on a positive note, they found that if our world stabilizes atmospheric carbon dioxide at or below 450 parts per million, the polar bears might survive throughout much of their current range. Enough arctic ice would likely remain during late summer and early autumn if nations increase their carbon dioxide–reduction targets in the coming decades.[8]

In the words of longtime polar bear research biologist Steve Amstrup, "We have now shown that there is something that can be done to save the polar bears. This problem is not irreversible."

Let us collectively act upon these words.

Coupled with the challenges of global warming, the Arctic ironically serves as a taproot for fueling greenhouse gas emissions. America's largest network of oil fields is on the doorstep of the Arctic Refuge coastal plain. Each year the state of Alaska leases millions of acres of North Slope lands to the oil industry. While our nation and the world should be reducing consumption of fossil fuels, oil development is the lifeblood industry of Alaska's economy. Oil seekers and their political allies will continue to bang on the Arctic Refuge door of wishful bonanzas.

For the past three decades, the conservation community and the Gwich'in people have repeatedly defended the Arctic National Wildlife Refuge, advocating for wilderness protection of the fragile coastal plain. Those collective voices have halted numerous oil drilling schemes that would invade the birthplace of the polar bears, caribou, and a multitude of migratory birds from all corners of the world.

This unique, arctic birthplace *is* a sacred place.

This birthplace is located in one of the greatest wilderness areas on Earth. Oil development does not belong there.

Under international treaty obligations, the United States should take leadership in protecting this vital birthplace of our threatened polar bears, the Porcupine caribou herd, and the millions of migratory birds that know no boundaries. We share these wandering species with Canada and many other countries that provide winter homes for birds that begin their lives in the Arctic.

Given the magnitude of climate change, and the global movement toward a clean energy future, this is the perfect time for both the president and Congress to take action. The president can demonstrate great international leadership by declaring the coastal plain an Arctic Wildlife Monument, in support of the denning polar bears, calving caribou, and nesting birds. Congress in turn should pass companion legislation designating the coastal plain a wilderness area, giving it the highest level of protection possible under current law.

Let's raise our voices together.

—Debbie Miller, 2011

Notes

1. Old-squaw ducks have been renamed long-tailed ducks (*Clangula hyemalis*).
2. According to a current (2011) USFWS biologist, there are now forty-five bird species that utilize the lagoons, mudflats, and shoreline of the Arctic Refuge.
3. Tragically, the 2010 Deepwater Horizon oil spill disaster was much larger than the 1989 Exxon Valdez spill. An estimated 5 million barrels of oil gushed into the Gulf of Mexico over a three-month period, making it the largest marine spill in history. In May 2010, the surface oil slick measured 29,000 square miles, about the size of South Carolina or roughly the same size as the Arctic Refuge. As in the case of the Exxon Valdez spill, it will take decades for the marine environment and the people who live off the sea to fully recover.
4. The lesser golden plover was divided into two species in 1993: the American golden-plover (*Pluvialus dominica*) and the Pacific golden-plover (*Pluvialus fulva*). Most American golden-plovers spend the winter in South America, while the majority of Pacific golden-plovers migrate to Hawaii and other islands in the South Pacific. Their summer breeding ranges largely overlap in Alaska. The American golden-plover is more commonly seen in the Arctic Refuge.
5. Studies as of 2010 show that there is an increasing trend for polar bears to den on land because of deteriorating sea ice conditions. Two-thirds of the Beaufort Sea pregnant bears now choose to den on land. The Beaufort Sea population has decreased to 1,500 bears. Refer to the new afterword in this book for additional information regarding the worldwide status of polar bears and the effects of global warming.
6. No longer endangered, the peregrine falcon has made a strong recovery. In the Arctic Refuge, peregrine nests have steadily increased in numbers along the Porcupine River, from five nests in 1976 to as many as thirty nests in 2010, according to USFWS researchers.
7. Department of Environment and Natural Resources, Northwest Territories, *Species at Risk in the Northwest Territories* (Yellowknife, NT: Government of the Northwest Territories, 2010).
8. Steven C. Amstrup, Eric T. DeWeaver, Bruce G. Marcot, et al., "Greenhouse Gas Mitigation Can Reduce Sea-Ice Loss and Increase Polar Bear Persistence," *Nature* 468 (December 16, 2010), 955-58.

Bibliography

Alaska Department of Environmental Conservation, Spill Response Office. Personal Communication. Fairbanks, 1999.

Alaska Department of Natural Resources. *Five-Year Oil and Gas Leasing Program*. Fifteenth Alaska Legislature, Second Session. Anchorage: Division of Oil and Gas, 1988.

———. *Five-Year Oil and Gas Leasing Program*. Anchorage: Division of Oil and Gas, 1999.

———. *Historical and Projected Oil and Gas Consumption*. Anchorage: Division of Oil and Gas, 1999.

Amstrup, Steve. *Polar Bear*. Audubon Wildlife Report. New York: National Audubon Society, 1986.

Amstrup, Steven C., Eric T. DeWeaver, Bruce G. Marcot, et al. "Greenhouse Gas Mitigation Can Reduce Sea-Ice Loss and Increase Polar Bear Persistence." *Nature* 468, 955–958 (December 16, 2010).

Bland, John. *Forests of Lilliput: The Realm of Mosses and Lichens*. Series in Nature and Natural History. Englewood Cliffs, New Jersey: Prentice-Hall, 1971.

Brower, Kenneth. *Earth and the Great Weather: The Brooks Range*. Friends of the Earth. New York: McCall Publishing Company, 1973.

Calef, George. *Caribou and the Barren-Lands*. Canadian Arctic Resources Committee. Ottawa: Firefly Books Ltd., 1981.

Caulfield, Richard A. *Subsistence Land Use in Upper Yukon–Porcupine Communities, Alaska*. Technical Paper No. 16. Fairbanks: Alaska Department of Fish and Game, Division of Subsistence, 1983.

Chapman, Joseph A., and George A. Feldhamer, eds. *Wild Mammals of North America: Biology, Management, and Economics*. Baltimore: Johns Hopkins University Press, 1982.

Douglas, William O. *My Wilderness*. New York: Doubleday, 1960.

Franklin, Sir John. *Narrative of Journey to the Shores of the Polar Sea, in 1825-26-27*. London: J. Murray, 1829.

Hadleigh-West, Frederick. *The Netsi Kutchin: An Essay in Human Ecology*. Ann Arbor, Michigan: University Microfilms International, 1963.

Harms, Cathy, and Chris Smith. *Wolves Around the World: Symposium Studies International Status*, volume 20, number 6 (November-December 1988), pg. 25. Alaska Department of Fish and Game, 1988.

Irving, Laurence. *Arctic Life of Birds and Mammals, Including Man*. New York: Springer Verlag, 1972.

Jacobson, Michael J., and Cynthia Wentworth. *Kaktovik Subsistence: Land Use Values Through Time in the Arctic National Wildlife Refuge Area*. Fairbanks: U.S. Fish and Wildlife Service, Northern Ecological Services, 1982.

Leffingwell, Ernest de K. *The Canning River Region Northern Alaska*. U.S. Geological Survey Professional Paper 109. Washington, D.C.: Government Printing Office, 1919.

Leopold, A. Starker, and F. Fraser Darling. *Wildlife in Alaska*. New York: Ronald Press, 1953.

Libbey, David. *Kaktovik Area Cultural Resource Survey*. Cooperative Park Studies Unit, Anthropology and Historic Preservation. Fairbanks: University of Alaska, 1982.

Marshall, Robert. *Alaska Wilderness: Exploring the Central Brooks Range*. Berkeley: University of California Press, 1956.

Martin, Philip D., and Cathryn S. Moitoret. *Bird Populations and Habitat Use, Canning River Delta, Alaska*. Fairbanks: U.S. Fish and Wildlife Service, Arctic National Wildlife Refuge, 1981.

McCloskey, J. Michael, and Heather Spalding. *A Reconnaissance-Level Inventory of the Wilderness Remaining in the World*. Fourth World Wilderness Congress. Washington, D.C.: Sierra Club, 1987.

McKennan, Robert A. *The Chandalar Kutchin*. Technical Paper No. 17. Montreal: Arctic Institute of North America, 1965.

Muir, John. *My First Summer in the Sierra*. Boston: Houghton Mifflin Co., 1911.

Murie, Margaret E. *Two in the Far North*. Second edition. Edmonds,Washington: Alaska Northwest Publishing Company, 1978.

Naske, Claus M. "Creation of the Arctic National Wildlife Range," *National Wildlife Refuges of Alaska: A Historical Perspective*, Part 1. U.S. Department of Interior, U.S. Fish and Wildlife Service, 1979.

Nellemann, C., and Cameron, R. D. "Cumulative Impacts of an Evolving Oil Field Complex on the Distribution of Calving Caribou," *Canadian Journal of Zoology*, volume 76 (1998): 1425–1430, 1998.

Ritchie, Robert J., and Robert A. Childers. *Recreation, Aesthetics and Use of the Arctic National Wildlife Range and Adjacent Area, Northeastern Alaska*. Preliminary Report.

Sable, Edward G. *Geology of the Western Romanzof Mountains, Brooks Range, Northeastern Alaska*. U.S. Geological Survey Professional Paper No. 897. Washington, D.C.: Government Printing Office, 1977.

Speer, Lisa, and Sue Libenson. *Oil in the Arctic: The Environmental Record of Oil Development on Alaska's North Slope*. Washington, D.C.: Natural Resources Defense Council, 1988.

Stefansson, Vilhjalmur. *Discovery: The Autobiography of Vilhjalmur Stefansson*. New York: McGraw-Hill Book Company, 1964.

Trustees for Alaska. *Under the Influence: Oil and the Industrialization of America's Arctic*. Anchorage, 1998.

U.S. Congress, House Committee on Merchant Marine and Fisheries. *Hearings on the Arctic National Wildlife Refuge: Wildlife Issues* (H.R. 1082, H.R. 3601, H.R. 3928). 100th Congress, Second Session, 3 March 1988. Serial 100—52. Washington, D.C.: Government Printing Office, 1988.

———. Senate. Committee on Energy and Natural Resources. *Hearings on the Arctic Coastal Plain Public Lands Leasing Act of 1987, S. 1217*. 100th Congress, First Session, 15 October 1987. S. Hrg. 100–452. Washington, D.C.: Government Printing Office, 1987.

———. Senate. Committee on Interstate and Foreign Commerce. *Hearings on 8-1899, A Bill to Authorize the Establishment of the Arctic Wildlife Range, Alaska, and for Other Purposes*. 86th Congress, 29 October 1959. Fairbanks and Washington, D.C.: Government Printing Office, 1960.

U.S. Department of Defense. *1986 Final Report for the Endicott Environmental Monitoring Program*. Army Corps of Engineers draft report, 1986.

U.S. Department of Interior, U.S. Fish and Wildlife Service. *Arctic National Wildlife Refuge, Alaska, Coastal Plain Resource Assessment: Report and Recommendation to the Congress of the United States and Final Legislative Environmental Impact Statement.* Prepared in cooperation with U.S. Geological Survey and the Bureau of Land Management, 1987.

———. Public hearing records on *Acquisition of Selected Inholdings in Alaska's National Wildlife Refuges, Legislative Environmental Impact Statement*. Testimony from Kaktovik, Alaska, 27 September 1988, and Arctic Village, Alaska, 28 September 1988.

———. *Comparison of Actual and Predicted Impacts of the Trans-Alaska Pipeline System and Prudhoe Bay Oilfields on the North Slope of Alaska*. Draft report. Fairbanks: Fish and Wildlife Enhancement Office, 1987.

U.S. Environmental Protection Agency. *Report to Congress: Management of Wastes from the Exploration, Development, and Production of Crude Oil, Natural Gas, and Geothermal Energy*, volume 1 of *Oil and Gas*. Washington, D.C.: Office of Solid Waste and Emergency Response, 1987.

Warbelow, Cyndie, David Roseneau, and Peter Stern. "The Kutchin Caribou Fences of Northeastern Alaska and the Northern Yukon," chapter 4 in *Arctic Gas: Biological Report Series*, volume 32 of *Studies of Large Mammals along the Proposed MacKenzie Gas Pipeline Route from Alaska to British Columbia*. Edited by R. D. Jakimchuk, Renewable Resources Consulting Services Ltd., 1975.

West, George C., and David W. Norton. "Metabolic Adaptations of Tundra Birds," *Physiological Adaptations to the Environment*, edited by F. J. Vernberg. New York: Intext Educational Publishers, 1975.

West, Robin L., and Elaine Snyder-Conn. *Effects of Prudhoe Bay Reserve Pit Fluids on Water Quality and Macroinvertebrates of Arctic Tundra Ponds in Alaska*. Biological Report 87 (7). U.S. Department of Interior, U.S. Fish and Wildlife Service, 1987.

Arctic National Wildlife Refuge Contact Information

For more information about the Arctic National Wildlife Refuge contact:

Refuge Manager
Arctic National Wildlife Refuge
101 Twelfth Avenue, Room 236
Fairbanks, AK 99701
(800) 362-4546
email: arctic_refuge@fws.gov
http://arctic.fws.gov

The following conservation organizations are working to protect the Arctic National Wildlife Refuge. Some of these national groups may have a field office in your home state. For current information, contact any of the following:

Alaska Wilderness League
122 C Street, Northwest
Washington, DC 20001
(202) 544-5205
www.alaskawild.org

National Audubon Society
700 Broadway
New York, NY 10003
(212) 979-3000
www.audubon.org

National Wildlife Federation
11100 Wildlife Center Dr.
Reston, VA 20190
(800) 822-9919
www.nwf.org

Sierra Club
85 Second Street, 2nd Floor
San Francisco, CA 94105-3441
(415) 977-5500
www.sierraclub.org

The Wilderness Society
1615 M Street, Northwest
Washington, DC 20036
(800) 843-9453
www.wilderness.org

In Alaska contact:

Northern Alaska Environmental Center
830 College Road
Fairbanks, AK 99701
(907) 452-5021
www.northern.org

Index

L

M

O

Y

Z

About the Author

Debbie S. Miller grew up in the San Francisco Bay area near the redwood groves. In her youth, she spent much of her time exploring the natural world and hiking through the Sierra Nevada Mountains. At an early age she developed a love for writing, inspired by such nature writers as John Muir, Rachel Carlson, and William O. Douglas.

An explorer at heart, Debbie attended college near the Rocky Mountains where she enjoyed discovering the beauty of this mountain world. She received her B.A. in Education and teaching certification from the University of Denver in 1973.

In 1975, Debbie moved to Alaska with her husband, Dennis. They were both drawn to Alaska's wilderness out of their love for wild places. As teachers, the Millers lived in Arctic Village, a small Athabaskan Gwich'in Indian village located on the southern boundary of the Arctic National Wildlife Refuge. They were introduced to an extraordinary culture and a people closely connected to the wilderness and wildlife of their homeland. Debbie also had the opportunity to begin exploring and writing about the Arctic National Wildlife Refuge, one of the greatest wild regions remaining on Earth.

Over the past three decades, Debbie has authored many books, essays, and articles about Alaska and the magical world of the Arctic. Each year she continues to hike, float, climb, camp, study, and write about Alaska's rich diversity of environments and wildlife. Of all places, she cherishes the Arctic National Wildlife Refuge, which has been the greatest source of inspiration for her writing. *Midnight Wilderness* is based on fourteen years of explorations and journals that reflect her many trips through the Refuge.

Miller has also traveled to the remote Eskimo village of Nuiqsut to document the impacts of oil development on the Inupiat culture. Her in-depth article, "Ground Zero," was published in the Summer 2001 issue of the *Amicus Journal* and was nominated for a National Press Award in environmental reporting. In 2003, Miller's essay "Clinging to an Arctic Homeland" was published in the award-winning photographic book by Subhankar Banerjee, *Arctic National Wildlife Refuge: Seasons of Life and Land* (The Mountaineers Books). This essay offers two cultural perspectives on life in the Inupiat village of Kakovik, and in the Gwich'in community of Arctic Village. In another recent essay, Miller describes the fascinating lives of songbirds that migrate to the Arctic Refuge from five continents. "Songs from Around the World" is based on Miller's 2005 trip to the Arctic Refuge where she hiked seventy-five miles through the Sadlerochit Mountains and across the coastal plain, studying migrants such as the American Dipper, an amazing aquatic songbird that lives and swims in the Arctic Refuge year-round. This essay was published in *Arctic Wings: Birds of the Arctic National Wildlife Refuge* (Braided River, 2006).

Debbie is also the author of many award-winning nature books for children, illustrated by wildlife artists Jon and Daniel Van Zyle. Her books, such as *Survival at 40 Below, A Caribou Journey, Flight of the Golden Plover, River of Life*, and *Big Alaska: Journey Across America's Most Amazing State* have been recognized as Outstanding Science Trade Books for Children by the National Science Teachers Association. Her book, *Arctic Lights, Arctic Nights*, received the 2003 John Burroughs Nature book for Young Readers Award.

Several years ago Debbie received the Refuge Hero Award from the U.S. Fish and Wildlife Service for her nature writing and education and conservation work. She has worked to protect the Arctic Refuge from industrial development for two decades, and is a founding board member of the Alaska Wilderness League. She has testified three times before Congress in an effort to educate members about the extraordinary values of the Arctic Refuge.

Braided River, the conservation imprint of The Mountaineers Books, combines photography and writing to bring a fresh perspective to key environmental issues facing western North America's wildest places. Our books reach beyond the printed page as we take these distinctive voices and vision to a wider audience through lectures, exhibits, and multimedia events. Our goal is to inspire and motivate people to support critical conservation efforts and make a definitive difference.

We help shape the conversation about the importance of preserving wild places. We have reached more than 50,000 people through our outreach presentations and exhibits—and millions more through print and digital media. By developing partnerships with like-minded organizations and building broader public support, we inspire on-the-ground victories for wilderness preservation.

Please visit www.BraidedRiver.org for more information on events, exhibits, speakers, and how to contribute to this work.

Braided River books may be purchased for corporate, educational, or other promotional sales. For special discounts and information, contact our sales department at (800) 553-4453 or mbooks@mountaineersbooks.org.

Alaska Wilderness League leads the effort to preserve Alaska's wilderness by engaging citizens, sharing resources, collaborating with other organizations, educating the public, and providing a courageous, constant, and victorious voice for Alaska in the nation's capital.

About Us: Alaska's wild lands—including the Arctic National Wildlife Refuge—are under attack. This extraordinary treasure trove of lands, set aside decades ago to be protected now and in the future for the benefit of the American people, is in severe danger of being destroyed forever by short-sighted politicians and the extractive industries. They want only the resources these pristine areas can provide, regardless of the resulting devastation to the habitat, wildlife, and cultures.

Alaska Wilderness League is a non-profit 501(c)(3) corporation founded in 1993 to further the protection of Alaska's amazing public lands. The League is the only Washington, D.C.-based environmental group devoted full-time to protecting the Arctic National Wildlife Refuge and other wilderness-quality lands in Alaska.

Our Work: Alaska Wilderness League works at the federal level on a variety of issues affecting Alaska's wild land and waters. Currently, the League is fighting to permanently protect the Arctic National Wildlife Refuge as wilderness, promote the sustainable future of the Tongass National Forest, and check the unbalanced and potentially destructive development of Alaska's Arctic waters, Western Arctic, and other public lands.

Alaska Wilderness League, based in Washington, D.C., has offices in Juneau, Fairbanks, and an Arctic Environmental Justice Center in Anchorage, Alaska. The center provides a base of outreach and support for members of Arctic communities who are on the front lines of the destruction from industrial development.

ARCTIC WINGS
Birds of the Arctic National Wildlife Refuge
Stephen Brown
A tribute to the birds that journey to the refuge and back every year

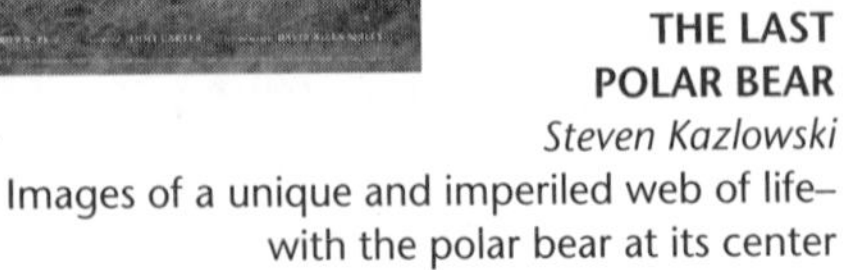

THE LAST POLAR BEAR
Steven Kazlowski
Images of a unique and imperiled web of life–with the polar bear at its center

ARCTIC NATIONAL WILDLIFE REFUGE
Seasons of Life and Land
Subhankar Banerjee
A comprehensive portrait of the refuge in fall, winter, spring, and summer

PLANET ICE
A Climate for Change
James Martin
Documents the power of ice and its unique role in revealing the condition of our planet

ICE BEAR
The Arctic World of Polar Bears
Steven Kazlowski
Reveals the polar bear's world, as well as the entire arctic landscape